SpringerBriefs in Molecular Science

Green Chemistry for Sustainability

Series editor

Sanjay K. Sharma, Jaipur, India

More information about this series at http://www.springer.com/series/10045

Muthupandian Ashokkumar

Ultrasonic Synthesis of Functional Materials

 Springer

Muthupandian Ashokkumar
School of Chemistry
University of Melbourne
Parkville, VIC
Australia

ISSN 2191-5407   ISSN 2191-5415   (electronic)
SpringerBriefs in Molecular Science
ISSN 2212-9898
SpringerBriefs in Green Chemistry for Sustainability
ISBN 978-3-319-28972-4   ISBN 978-3-319-28974-8   (eBook)
DOI 10.1007/978-3-319-28974-8

Library of Congress Control Number: 2015960231

Printed on acid-free paper

This Springer imprint is published by SpringerNature
The registered company is Springer International Publishing AG Switzerland

Preface

Ultrasonic technology has been increasingly used in synthetic applications despite its existence over a century for traditional industrial applications such as cleaning, extraction and emulsification. Over the past few decades, the potential use of ultrasound in synthetic reactions has been explored in laboratory-scale experiments. Due to unique and extreme reaction environment generated within cavitation bubbles, in otherwise a room temperature liquid, a variety of functional materials could be synthesised.

A number of books and review articles are available in the literature highlighting the benefits of ultrasound-driven chemical reactions including synthesis of materials. However, a textbook-style Brief on the use of ultrasound for synthesising functional materials is not available. As discussed in this Brief, functional materials are used by the human society for various purposes that include catalysis, energy conversion, biomedical applications, nutritional foods, etc. Ultrasonic synthetic methodology offers a relatively simple and green route to synthesise such functional materials.

Chapter 1 of this Brief provides fundamental science involved in ultrasonic synthesis of functional materials. How sound waves interact with gas bubbles in liquids and what are the consequences of such interactions that lead to the generation of a variety of chemical reactions are discussed in this chapter. Specific examples of ultrasonic synthesis of functional materials that include inorganic and organic nanomaterials, core–shell microspheres and functional foods have been highlighted in Chap. 2. The advantages, disadvantages and challenges of ultrasonic technology are briefed in Chap. 3.

The Brief on ultrasonic synthesis of functional materials will benefit wider academic and student communities as it provides a snapshot of both fundamental and applied aspects of past and recent developments in this developing research area. In addition, it may also benefit various industry sectors willing to embrace a novel technology in materials synthesis and food processing.

Muthupandian Ashokkumar

Acknowledgments

The author would like to acknowledge his collaborators, colleagues, postdoctoral scientists and students for their contributions to expand the knowledge base of ultrasonic technology. Most of the examples and data discussed in this Brief are the outcomes of collaborative work involving several people. The author would also like to acknowledge the Australian Research Council for funding several research projects under the Discovery Project, Linkage Project and Industry Transformation Research Hub programmes.

Contents

Chapter 1
Introduction

Abstract The fundamental science responsible for chemical and physical effects caused by ultrasound in a liquid medium is discussed in this chapter. Various events that occur when sound waves of appropriate frequency and power interact with a liquid medium are explained. Acoustic cavitation process and the generation of strong physical forces and highly reactive radicals have been described in simple terms. Also, the effect of acoustic frequency on the physical and chemical effects is discussed. Overall, this chapter provides a simplistic view of acoustic cavitation and associated events that is required to fully understand the processes discussed in Chap. 2.

Keywords Ultrasound · Acoustic cavitation · Cavitation bubbles · Ultrasound frequency · Bubble temperature · Sonochemistry

Functional materials are integral part of human life [1, 2]. Considering the health of human beings, drug delivery vehicles and materials possessing antimicrobial activities and nutritional properties come under this category. Delivering nutritional compounds and pharmaceuticals through commonly consumed food matrix is in practice. For example, non-polar nutrients are dispersed as emulsions in water-based drinks including milk [3]. Similarly, inorganic and organic nano- and micro materials with functional properties are increasingly being used in various applications. For example, metal and semiconductor nanoparticles are used as catalytic materials [4, 5].

Novel methodologies for synthesising functional materials have been constantly developed. Nanomaterials for energy conversion and biological applications could be synthesised by sol-gel methods and conventional chemical reactions to precisely control their size and size distribution [6–10]. Layer-by-layer method to generate core-shell materials for drug delivery applications is a well-known procedure [11] that primarily works on electrostatic or hydrophobic interactions between functional molecules. Emulsion polymerisation is another commonly used method to generate core-shell functional materials [12–14]. Strong shear forces are applied and surfactant/polymer based stabilisers are used to stabilise for example an oil-in-water

© The Author(s) 2016
M. Ashokkumar, *Ultrasonic Synthesis of Functional Materials*,
SpringerBriefs in Green Chemistry for Sustainability,
DOI 10.1007/978-3-319-28974-8_1

emulsion droplet. The emulsion droplet itself could be a nutrient dispersed in a food or drink matrix [15]. Such emulsions are also used to generate functional polymer latex particles [16].

Ultrasonic synthesis of functional materials has been widely reported and is a growing methodology that has significant potential to generate large-scale functional materials. This Brief is aimed at providing the fundamental science involved in ultrasonic synthesis of functional materials with specific examples. The advantages, disadvantages and challenges of this technology have also been highlighted.

1.1 Sound

Animal kingdom communicate with each other using soundwaves. Soundwaves can be divided into various frequency range based on their uses/applications [17]. Accordingly, below 20 Hz frequency range is referred as subsonic or infrasonic waves. Human ear cannot detect this frequency range. Sometimes, such frequency range can be felt as shock waves, for example during earth quakes. About 20 Hz–20 kHz is in the range of human hearing. Soundwaves above 20 kHz frequency are referred as ultrasound, which are subdivided into power ultrasound, used in a variety of physical and chemical applications and magasonic, above 1 MHz, used in medical diagnostics. Most of the studies dealing with material synthesis use 20 kHz–1 MHz frequency range.

The speed of sound varies depending upon the medium. In air, sound travels at a speed of about 320 m/s and its speed in water is 1500 m/s. Sound travels relatively faster in solids—for example, the speed of sound in steel is about 5000 m/s. The mechanical vibrations generated by sound waves have been used in various applications [18, 19]. For example, rate of dehydration of specific materials could be enhanced by ultrasonic vibration [20]. Acoustic reflection and scattering techniques have been used in quality control of food and other materials [21, 22]. Ultrasonic imaging is a well-known technique in diagnostic medicine [23]. Ultrasonic cleaning of small and large equipment is a well-known industrial application [24]. Ultrasonic cleaning is also a common practice in dental clinics [25]. Ultrasonic extraction and emulsification are used in food and pharmaceutical industries [26–30].

While soundwaves are commonly used for communication through various media, some unique events occur in liquids when soundwaves interact with the medium. In particular, when ultrasound passes through a liquid medium, it strongly interacts with small gas bubbles that exist in the liquid. Such interaction between ultrasound and gas bubbles leads to a phenomenon known as acoustic cavitation [31, 32].

1.2 Acoustic Cavitation

Cavitation literally means creation of a cavity. Hence, acoustic cavitation literally refers to the formation of a cavity using acoustic (sound) energy. Such a process requires significant amount of energy. For example, to separate water molecules apart overcoming intermolecular forces to a distance in the order of nanometres requires a negative pressure of hundreds of atmospheres [17]. Equation 1.1 can be used to calculate the critical pressure (P_B) required to create a bubble of radius, R_e.

$$P_B \sim P_h + \frac{0.77\sigma}{R_e} \tag{1.1}$$

σ is surface tension of the liquid and P_h is hydrostatic pressure (could be approximated to atmospheric pressure under normal experimental conditions). Note that this equation is valid when $2\sigma/R_e \ll P_h$.

However, cavities could be formed from pre-existing gas nuclei at much lower power levels [33–35]. Gas nuclei, inherently present in liquids, are forced to oscillate under the fluctuating pressure field when ultrasound passes through a liquid. Gas bubbles formed from such oscillations grow over a period by a process known as rectified diffusion [36, 37]. Bubble growth by rectified diffusion can be understood by area and shell effects. During the growth phase of a bubble, solvent vapour and dissolved gas molecules diffuse into the bubble. Since the surface area of the bubble wall increases during the growth phase, more molecules diffuse into the bubble. The opposite process occurs during the compression phase—molecules diffuse out of the bubble. However, the surface area decreases in the compression process and hence a relatively lower amount of molecules diffuse out of the bubble. For each expansion/compression cycle, more vapour/gas molecules stay inside the bubble resulting in a net growth of the bubble size. Another aspect that needs to be considered is the time scale involved for the expansion and compression phases. While an acoustic cycle has equal time for rarefaction and compression half cycles, single bubble dynamics experiments have shown that the growth phase of a bubble is a relatively slow process resulting in a higher amount of gas/vapour molecules diffusing into a bubble. In addition, dissolved gas concentration in the liquid shell surrounding an oscillating bubble varies significantly. In the compressed state of a bubble, gas molecules diffuse out of the bubble into the liquid shell surrounding a bubble. The diffusion of gas molecules becomes relatively slow and hence less gas molecules diffuse out of the bubble. The process works favourably during expansion process. Gas molecules can freely diffuse into the bubble. The shell effect is a very complex process than what is described above. A combination of area and shell effects and the time scale involved in these processes lead to rectified diffusion growth of bubbles in an acoustic field. Detailed models have been developed to theoretically estimate the growth of bubbles by rectified diffusion process [38].

Bubble growth by rectified diffusion is not an infinite process. The growth of the bubble is restricted by the applied frequency. The wall of a free gas bubble in a liquid oscillates at a given frequency depending upon the size of the bubble. The relationship between oscillation frequency and radius of a bubble is given by a simplified form of the Minnaert's Equation (Eq. 1.2) [39].

$$f * R \sim 3 \tag{1.2}$$

where, f is frequency in Hz (s^{-1}) and R is resonance radius of the bubble in m. For example, the wall of a 150 µm radius bubble oscillates at a resonance frequency of 20 kHz.

Note that despite a single resonance size is expected theoretically for a given frequency, a size range exists [40]. Using pulsed sonoluminescence and sonochemiluminescence methods, the resonance size range of SL and sonochemically active bubbles have been estimated [41, 42]. Lee et al. developed a novel pulsed sonoluminescence (SL) technique [41] to determine the resonance size range of cavitation bubbles. By systematically increasing the pulse off time, they could see a decrease in the SL intensity due to the dissolution of active cavitation bubbles below the resonance size during this period. Beyond a certain pulse off time, no more SL intensity could be observed indicating the dissolution of all cavitation bubbles below the resonance size. This is schematically shown in Fig. 1.1.

Taking the pulse off time at which SL disappeared, the resonance size of the cavitation bubbles was calculated. Epstein Plesset equation, (Eq. 1.3) which relates bubble dissolution time to its radius, was used for this purpose.

$$\left(\frac{DC_s}{\rho_g R_o^2}\right)t = \frac{1}{3}\left(\frac{RT\rho_g R_o}{2M\gamma}\right) + 1 \tag{1.3}$$

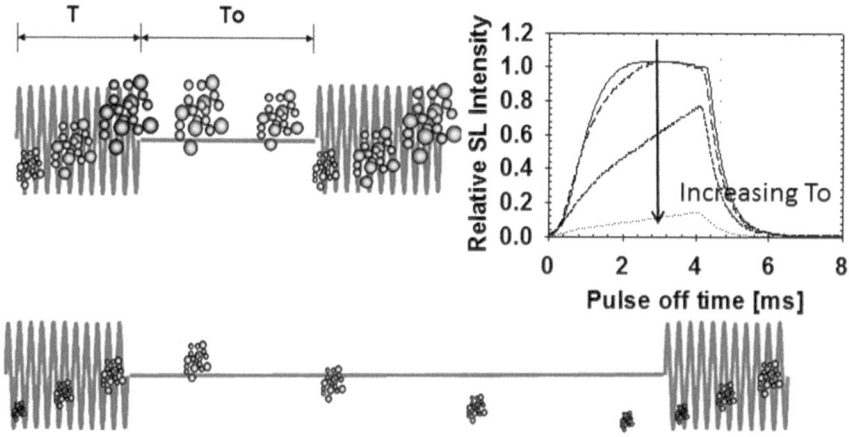

Fig. 1.1 Schematic representation [Adapted from Ref. 41] of pulsed sonoluminescence technique to determine the resonance size of cavitation bubbles. Bubbles grow during pulse on (T) and dissolve during pulse off (T_o). With increasing T_o, steady-state SL intensity decreases (*top right*) eventually to zero—the corresponding To is used to calculate the bubble size using Eq. 1.3

D—diffusion coefficient; C_s—dissolved gas concentration; ρ_g—density of gas; R_o—initial bubble radius; t—dissolution time; M—molar mass of gas; γ—surface tension of the liquid; R—gas constant and T—solution temperature. Replacing t with T_o, R_o (assumed to be equal to resonance size) can be determined. The resonance size of cavitation bubbles at 515 kHz was found to be in the range 2.8–3.8 μm.

A follow up work by Brotchie et al. [42] noted that the resonance sizes of sonoluminescence and sonochemically active bubbles are different. The sonochemiluminescence, resulting from the reaction between OH radicals generated within cavitation bubbles and luminol molecules, intensity was used to determine the sonochemically active (SCL) bubbles. The resonance size of SL bubbles are found to be relatively larger than that of SCL bubbles. In addition, Eq. (1.2) shows that the resonance size decreases with an increase in ultrasonic frequency. Brotchie et al. [42] have also confirmed this experimentally. The sizes were found to be 3.9, 3.2, 2.9, 2.7 and 2 μm at 213, 355, 647, 875 and 1056 kHz frequency, respectively. Another important aspect that needs to be mentioned is the difference between theoretical and experimentally determined resonance sizes of the cavitation bubbles. Equation (1.2) provides a theoretical value of 14 μm at 213 kHz whereas the experimental value is found to be 3.9 μm. This is also known from single bubble work at 20 kHz where the experimental resonance size was found to be about 5 μm compared to the theoretical value of 150 μm [43]. The difference between the resonance size determined by Eq. (1.2) and experimental value is due to the fact that Eq. (1.2) is a very simplified one that does not consider the physical properties of the liquid or bubble contents.

When bubbles reach the resonance size range, they grow to a maximum size within one acoustic cycle and implode. Bubble implosion/collapse is a near adiabatic process. In simple thermodynamic terms, the volume of the bubble decreases instantaneously resulting in the generation of extreme heat within the bubble. Theoretical estimates predict greater than 15,000 K [44, 45]. However, experimental methods estimate about 1000–5000 K [46–50]. A number of techniques have been used to calculate the bubble temperatures. First, the bubble temperature could be theoretically calculated using Eq. (1.4).

$$T_{max} = T_m \left(\frac{P_m(\gamma - 1)}{P_v} \right) \tag{1.4}$$

T_{max}—bubble temperature on collapse; T_m and P_m are solution temperature and pressure, respectively, P_v is pressure inside the bubble and γ is heat capacity ratio of the gas inside the bubble. A theoretical temperature of about 12,700 K could be calculated by using $\gamma = 1.66$ (ideal gas), $T_m = 298$ K, $P_m = 2$ atm, $P_v = 0.031$ atm. Replacing γ of an ideal gas by that of water (1.32), the temperature drops to 6150 K highlighting the importance of the heat capacity ratio of the gas contained in the collapsing bubbles. Suslick and coworkers [46, 47] have used sonoluminescence spectra to calculate bubble temperatures in multibubble systems and found to be in the order of 1000–5000 K. Henglein and coworkers [48] have used methyl radical recombination method and determined the cavitation bubble temperatures to be in a similar range. Independent of the accuracy of these methods, it could be realised that the extreme temperature conditions are generated within cavitation bubbles.

1.3 Physical and Chemical Effects of Acoustic Cavitation

When bubble collapse occurs, a number of physical and chemical events are generated. High intensity shock waves are generated when bubbles collapse symmetrically [51, 52]. The energy associated with shock waves are extremely high that could be used to increase mass transfer processes in liquids. Suslick and coworkers have shown [53] that shockwaves could drive particle-particle collision generating temperatures as high as 3500 K on the surface of the colliding particles. Another significant force that is generated during the asymmetric collapse of a cavitation bubble is microjet [54]. When a bubble is near a boundary, it experiences uneven acoustic force around it and undergoes asymmetric collapse with a liquid jet rushing towards the middle of the bubble hitting the surface/boundary with a speed of greater than 100 miles per second. Such high-speed jets possess enormous kinetic energy that could make pits or holes on the surface of a metal plate or particle. Other physical forces that are generated during acoustic cavitation are microstreaming, agitation, turbulence, etc. that could be used to enhance mass transfer effects in a medium [55]. Ultrasonic synthesis of polymer latex particles and core-shell biomaterials need strong shear forces and a small amount of radicals to initiate polymerisation or cross-linking reactions. 20 kHz horn system is found to be suitable for such applications [56–60].

The generation of very high temperatures on bubble collapse has already been mentioned. The heat generated can raise the temperature of the core of the bubbles to thousands of degrees for a short period (micro- to nanoseconds). Such extreme thermal conditions lead to light emission from the bubbles, referred to as sonoluminescence [61], which will not be discussed in this book. Another consequence of the high temperature conditions within the core of the cavitation bubble is the induction of a variety of chemical reactions [62–70]. In organic solvents, the heat generated has been used to synthesise amorphous metal nanoparticles (see Sect. 2.2). In aqueous solutions, water and other volatile compounds could diffuse into the bubble and pyrolysed/decomposed by the extreme temperatures of the bubble. If it is pure water saturated with argon, only **H** and **OH** radicals are predominantly generated and are referred to as primary radicals (Reaction 1).

$$\mathbf{H_2O} \rightarrow \mathbf{H} + \mathbf{OH} \qquad \text{(Reaction 1)}$$

The primary radicals may undergo recombination reactions to for molecular products (Reactions 2–4).

$$\mathbf{H} + \mathbf{OH} \rightarrow \mathbf{H_2O} \qquad \text{(Reaction 2)}$$

$$\mathbf{OH + OH \rightarrow H_2O_2} \qquad \text{(Reaction 3)}$$

$$\mathbf{H + H \rightarrow H_2} \qquad \text{(Reaction 4)}$$

In air-saturated water, a variety of radicals and molecular products are generated (Reactions 5–8)

$$\mathbf{H + O_2 \rightarrow HO_2} \qquad \text{(Reaction 5)}$$

$$\mathbf{HO_2 + HO_2 \rightarrow H_2O_2 + O_2} \qquad \text{(Reaction 6)}$$

$$O_2 \rightarrow 2O \qquad \text{(Reaction 7)}$$

$$O_2 + O \rightarrow O_3 \qquad \text{(Reaction 8)}$$

The reaction between N_2 and O_2 within cavitation bubbles has been shown to produce nitric acid, responsible for lowering the solution pH [71]. When an organic solute such as alcohol is dissolved in water, secondary reducing radicals are generated [72] (discussed in Sect. 2.1).

In simple terms, a variety of redox radicals are generated within the cavitation bubbles and hence each cavitation bubble could be compared to an electrochemical cell. For certain applications, only reducing radicals could be preferred: for example, for the reduction of metal ions to form metal nanoparticles (discussed in Sect. 2.2). In this case, the addition of a small amount of organic solute could convert all oxidising radicals (**OH** for example) to secondary reducing radicals. For reactions involving oxidation only, purging the solution with oxygen could convert the reducing radicals into oxidising radicals (Reaction 5).

The primary and secondary radicals could also be used in polymerisation and other chemical reactions [72–80]. Specific examples are discussed in Chap. 2. Prior to looking at such examples, it should be highlighted that the choice of right frequency is important to achieve optimal efficiency for every reaction. For this purpose, reactions/processes could be grouped into 3 categories: reactions that primarily depend upon the physical effects of ultrasound, reactions that primarily depend upon the chemical effects of ultrasound and reactions that depend upon both physical and chemical effects. For example, ultrasonic depolymerisation reactions, ultrasonic emulsification, ultrasonic extraction and ultrasonic cleaning primarily depend upon the physical forces (shear forces, microstreaming, microjetting, shock waves, agitation, etc.) generated during acoustic cavitation [81–85]. Ultrasonic synthesis of nanomaterials in aqueous solutions, sonochemical degradation of pollutants, etc., depend primarily on the amount of primary and secondary radicals generated during acoustic cavitation [86–90]. Ultrasonic synthesis of polymer latex particles and core-shell biomaterials depend upon both physical and chemical effects [91–95]. The mechanism behind such reactions/processes will be discussed later. However, it is worth noting that the physical forces generated at 20 kHz in a horn-type sonicator is significantly stronger. For this reason, applications such as extraction, cleaning and emulsification use a horn-type sonicator. The amount of radicals generated in such system is significantly lower compared to that generated at high frequencies where plate-type transducers are used.

At 20 kHz, most of the cavitation activity occurs at the tip of the horn and hence the number of active bubbles generated is relatively lower. Despite the amount of radicals generated per bubble is higher, the overall yield is lower. At 20 kHz, the heat generated within the bubble could be significantly higher compared to that generated at higher frequencies since the bubble size is larger. Equation 1.2 shows the relationship between resonance bubble size and frequency. Accordingly, the amount of radicals generated per bubble decreases with increasing frequency. On the other hand, the number of bubbles generated increases with an increase in frequency (for a given volume and power input) leading to an increase in the amount of primary and secondary radicals with an increase in frequency. The increase in number of active cavitation bubbles is due to an increase in the number of standing waves as schematically and photographically shown in Fig. 1.2.

Hence, for achieving redox reactions, higher frequencies are found useful. It should also be noted that the radical yield reaches a maximum level in the frequency range, 200–800 kHz beyond which the yield is found to decrease [96–100]. The amount of primary radicals generated during acoustic cavitation could be quantified using a relatively simple iodide oxidation process. **OH** radicals generated react among themselves to produce hydrogen peroxide (Reaction 2). In the absence of any additive (in pure water), the amount of H_2O_2 produced remain reasonably stable for a short period of time. However, in the presence of iodide ions, H_2O_2 could be used to oxidise iodide ions to molecular iodine, which is useful to quantify the amount of OH radicals generated [100].

Figure 1.3 shows that the amount of **OH** radicals generated is the highest at 355 kHz among the three frequencies investigated [100]. The decrease observed at the highest frequency is due to a relatively lower bubble temperature and a lower amount of water vapour that could evaporate into cavitation bubbles during the expansion phase (due to relatively less time available for rarefaction cycle) at very high frequencies. A detailed discussion on this is available elsewhere [101].

In brief, for a given solution volume and acoustic power, a change in acoustic frequency results in an increase in the number of active bubbles and a decrease in the resonance size of the bubble. This would have two opposing effects. A decrease in bubble size means a decrease in collapse intensity and hence lower bubble temperature. This leads to a decrease in the amount of primary and secondary radicals generated per bubble. In the meantime, an increase in the number of bubbles (due to an increase in the number of standing waves) leads to an increase in the amount of radicals generated. It has been shown in many studies [96–100] that the sonochemical reaction yield peaks around 200–800 kHz beyond which a decline in the yield is observed.

Despite an increase in the number of bubbles, a decrease in bubble temperature and very short time available for volatile molecules to diffuse into the bubbles during the expansion cycle leading to the observed decrease in radical yield. This has been theoretically demonstrated in Fig. 1.4. It is known that solvent molecules adsorb on the surface of cavitation bubbles. Using the resonance radius of the bubbles at each

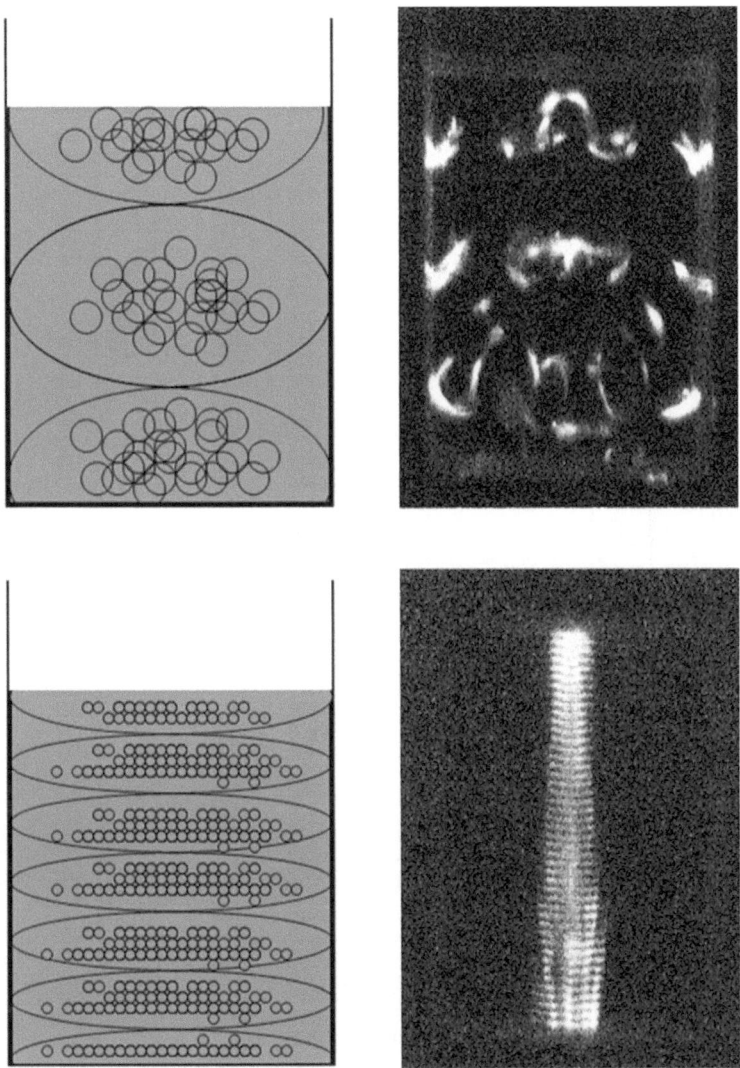

Fig. 1.2 *Left* Schematic representations of standing waves leading to an increase in the number of cavitation bubbles; *Right* photographic images of SL observed at 37 and 440 kHz

frequency, the amount of water molecules in a monolayer on the surface of bubbles could be calculated [101]. Also, there is a finite time required for the evaporation process to occur. Using the number of molecules at the interface and the time required for evaporation and expansion cycle, it could be shown that there is enough time available for the evaporation of one monolayer of water at low frequencies, whereas not enough time is available at high frequencies for the evaporation of one monolayer to be completed.

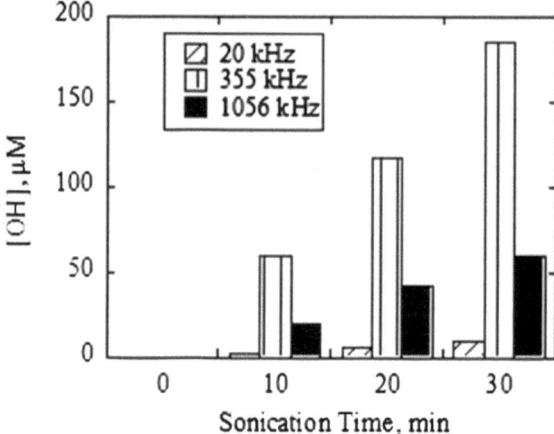

Fig. 1.3 OH radical yield as a function of sonication time at 20, 355 and 1056 kHz Adapted from Ref. [100]

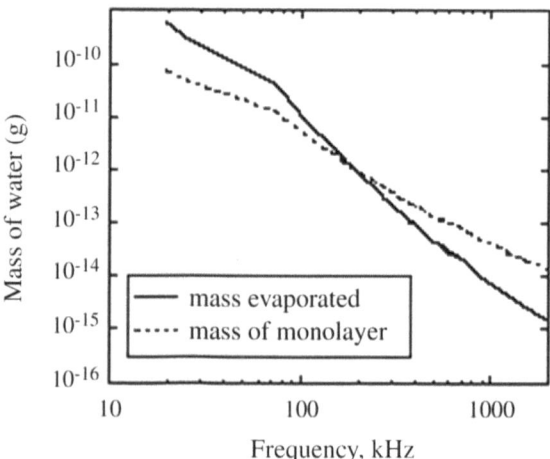

Fig. 1.4 Theoretical data shown mass of evaporated from bubble surface during a single expansion phase at various frequencies. It could be seen that the mass that could evaporate exceeds the amount present in a monolayer on the bubble surface at lower frequencies. At higher frequencies, the amount that could evaporate is less than a monolayer, which is due to very short expansion time available during bubble oscillations. Further details are available in Ref. [101]

It could be seen in Fig. 1.4 that the transition occurs around 200 kHz and gap widens at higher frequencies. However, the decrease in radical yield per bubble is compensated by the increase in number of bubbles until about 800 kHz as observed in many studies.

In summary, this chapter provides a simple overview of the various physical and chemical effects generated during acoustic cavitation and how different experimental parameters could be manipulated to control such effects.

References

1. D. Giugliano, Dietary antioxidants for cardiovascular prevention. Nutr. Metabol. Cardiovasc. Dis. **10**, 38–44 (2000)
2. M.R. Hoffmann, P.A. Senior, D.R. Mager, Vitamin D supplementation and health-related quality of life: a systematic review of the literature. J. Acad. Nut. Diet. **115**, 406–418 (2015)
3. H.M. Xing, L. Li, S.L. Gui, W. Elfallhe, S.H. He, Q.H. Sheng, Formation, stability, and properties of an algae oil emulsion for application in UHT milk. Food Bioprocess Technol. **7**, 567–574 (2014)
4. Md Rashid, M. Harunar, K. Tarum, Templateless synthesis of polygonal gold nanoparticles: an unsupported and reusable catalyst with superior activity. Adv. Funct. Mater. **18**, 2261–2271 (2008)
5. R. Velmurugan, M. Swaminathan, An efficient nanostructured ZnO for dye sensitized degradation of reactive red 120 dye under solar light. Sol. En. Mater. **95**, 942–950 (2011)
6. N. Pan, B. Wang, X.P. Wang, J.G. Hou, Manipulating and tailoring the properties of 0-D and 1-D nanomaterials. J. Mater. Chem. **20**, 5567–5581 (2010)
7. Z. Zhuang, Q. Peng, Y. Li, Controlled synthesis of semiconductor nanostructures in the liquid phase. Chem. Soc. Rev. **40**, 5492–5513 (2011)
8. J. Akbari, A. Heydari, Synthesis of Mn_3O_4 nanoparticles with controlled morphology using ionic liquid. Curr. Nanosci. **8**, 398–401 (2012)
9. T. Yu, A. Malugin, H. Ghandehari, Impact of silica nanoparticle design on cellular toxicity and hemolytic activity. ACS Nano **5**, 5717–5728 (2011)
10. R. Bohara, S.H. Pawar, Innovative developments in bacterial detection with magnetic nanoparticles. App. Biochem. Biotechnol. **176**, 1044–1058 (2015)
11. Y. Wang, A.S. Angelatos, F. Caruso, Template synthesis of nanostructured materials via layer-by-layer assembly. Chem. Mater. **20**, 848–858 (2008)
12. H.N. Yow, A.F. Routh, Formation of liquid core-polymer shell microcapsules. Soft Matter **2**, 940–949 (2006)
13. K. Li, X.R. Zeng, H.Q. Li, X.J. Lai, Role of acrylic acid in the synthesis of core-shell fluorine-containing polyacrylate latex with spherical and plum blossom-like morphology. J. App. Polym. Sci. **132** (2015) (Article number 42527)
14. Z. Ai, G. Sun, Q. Zhou, C. Xie, Polyacrylate-core/TiO_2-shell nanocomposite particles prepared by in situ emulsion polymerization. J. App. Polym. Sci. **102**, 1466–1470 (2006)
15. L. Gouveia, A. Raymundo, A.P. Batista, I. Sousa, J. Empis, *Chlorella vulgaris* and *Haematococcus pluvialis* biomass as colouring and antioxidant in food emulsions. Eur. Food Res. Technol. **222**, 362–367 (2006)
16. M.L. Wang, Z.J. Ma, D. Zhu, D.Y. Zhang, W. Yin, Core-shell latex synthesized by emulsion polymerization using an alkali-soluble resin as sole surfactant. J. App. Polym. Sci. **128**, 4224–4230 (2013)
17. T. Mason, J.P. Lorimer, *Sonochemistry: Theory, Applications and Uses of Ultrasound in Chemistry* (Wiley-Interscience, New York, 1989)
18. E. Verstrynge, M. Weavers, A novel technique for acoustic emission monitoring in civil structures with global fiber optic sensors. Smart Mater. Str. **23** Article Number: 065022 (2014)
19. K. Barron, Detection of fracture initiation in rock specimens by use of a simple ultrasonic listening device, international. J. Rock Mech. Mining Sci. **8**, 55–58 (1971)

20. M. Bantle, J. Hanssler, Ultrasonic convective drying kinetics of clipfish during the initial drying period. Dry. Technol. **31**, 1307–1316 (2013)
21. D.S. Morrison, U.R. Abeyratne, Ultrasonic technique for non-destructive quality evaluation of oranges. J. Food Eng. **141**, 107–112 (2014)
22. R. Bhaskaracharya, S. Kentish, M. Ashokkumar, Selected applications of ultrasonics in food processing. Food Eng. Rev. **1**, 31–49 (2009)
23. P.A. Payne, Y. Yerima, Extending medical ultrasound into new areas. Brit. J. Non-Destruct. Test. **33**, 7–10 (1991)
24. Anonymous, Multi stage ultrasonic cleaning system for the metal, electronics, aerospace and allied industries. Lubr. Eng. **52**, 669–669 (1996)
25. M. Vassey, C. Budge, T. Poolman, P. Jones, D. Perrett, N. Nayuni, P. Bennett, P. Groves, A. Smith, M. Fulford, P.D. Marsh, P.D.J.T. Walker, J.M. Sutton, N.D.H. Raven, A quantitative assessment of residual protein levels on dental instruments reprocessed by manual, ultrasonic and automated cleaning methods. Br. Dental J. **210**, 418–419 (2011)
26. T.J. Mason, L. Paniwnyk, F. Chemat, M.A. Vian, *Ultrasonic Food Processing*, ed. by A. Proctor. Alternatives to Conventional Food Processing, RSC Green Chemistry Series (2011) pp. 387–414
27. M.D. Esclapez, J.V. Garcia-Perez, A. Mulet, J.A. Carcel, Ultrasound-assisted extraction of natural products. Food Eng. Rev. **3**, 108–120 (2011)
28. Y. Shi, H. Li, J. Li, D.J. Zhi, X.Y. Zhang, H. Liu, H.Q. Wang, H.Y. Li, Development, optimization and evaluation of emodin loaded nanoemulsion prepared by ultrasonic emulsification. J. Drug Deliv. Sci. Technol. **27**, 46–55 (2015)
29. S. Chatterjee, M.A. Zaher, Encapsulation of fish oil with N-stearoyl O-butylglyceryl chitosan using membrane and ultrasonic emulsification processes. Carbohydr. Polym. **123**, 432–442 (2015)
30. M. Tabibiazar, A. Davara, M. Hashem, A. Homayonirad, F. Rasoulzadeh, H. Hamishehkar, M.A. Mohammadifar, Design and fabrication of a food-grade albumin-stabilized nanoemulsion. Food Hydrocoll. **44**, 220–228 (2015)
31. F.R. Young, *Cavitation* (Imperial College Press, 2000)
32. A.S. Peshkovsky, S. Peshkovsky, *Acoustic Cavitation Theory & Equipment Design Principles for Industrial Applications of High-Intensity Ultrasound* (Nova Science Publishers Inc., 2011)
33. Y.T. Didenko, T.V. Gordeychuk, Multibubble sonoluminescence spectra of water which resemble single-bubble sonoluminescence. Phys. Rev. Lett. **84**, 5640–5643 (2000)
34. T. Uchida, S. Takeuchi, T. Kikuchi, Measurement of amount of generated acoustic cavitation: investigation of spatial distribution of acoustic cavitation generation using broadband integrated voltage. Jpn. J. App. Phys. **50**, Article Number: 07HE01 (2011)
35. M. Lim, M. Ashokkumar, Y. Son, The effects of liquid height/volume, initial concentration of reactant and acoustic power on sonochemical oxidation. Ultrason. Sonochem. **21**, 1988–1993 (2014)
36. L.A. Crum, Acoustic cavitation series, 5. Rectified diffusion. Ultrasonics **2**, 215–223 (1984)
37. T. Leong, S. Wu, S. Kentish, M. Ashokkumar, Growth of bubbles by rectified diffusion in aqueous surfactant solutions. J. Phys. Chem. C **114**, 20141–20145 (2010)
38. L.A. Crum, G.M. Hansen, Generalized equations for rectified diffusion. J. Acoust. Soc. Am. **72**, 1586–1592 (1982)
39. M. Minnaert, On musical air-bubbles and the sound of running water. Philos. Mag. **16**, 235–248 (1933)
40. K. Yasui, Influence of ultrasonic frequency on multibubble sonoluminescence. J. Acoust. Soc. Am. **112**, 1405–1413 (2012)
41. J. Lee, M. Ashokkumar, S. Kentish, F. Grieser, Determination of size distribution of sonoluminescence bubbles in a pulsed acoustic field. J. Am. Chem. Soc. **127**, 16810–16811 (2005)
42. A. Brotchie, F. Grieser, M. Ashokkumar, The effect of power and frequency on acoustic cavitation bubble size distributions. Phys. Rev. Lett. **102**, 084302-1–084302-4 (2009)

43. M. Ashokkumar, L.A. Crum, C.A. Frensley, F. Grieser, T.J. Matula, W.B. McNamara III, K.S. Suslick, Effect of solutes on single-bubble sonoluminescence in water. J. Phys. Chem. A **104**, 8462–8465 (2000)

44. T. Leighton, *The Acoustic Bubble* (Academic Press, London, 1994)

45. K. Yasui, Single-bubble sonoluminescence from noble gases. Phys. Rev. E. **63**, Article Number: 035301 (2001)

46. Y.T. Didenko, W.B. McNamara, K.S. Suslick, Molecular emission from single-bubble sonoluminescence. Nature **407**, 877–879 (2000)

47. Y.T. Didenko, W.B. McNamara, K.S. Suslick, Sonoluminescence temperatures during multi-bubble cavitation. Nature **401**, 772–775 (1999)

48. J. Buttner, M. Gutierrez, A. Henglein, Sonolysis of water-methanol mixtures. J. Phys. Chem. **95**, 1528–1530 (1991)

49. A. Tauber, G. Mark, H.P. Schuchmann, C. von Sonntag, Sonolysis of tert-butyl alcohol in aqueous solution. J. Chem. Soc. Perkin Trans. **2**, 1129–1135 (1999)

50. J. Rae, M. Ashokkumar, O. Eulaerts, C. von Sonntag, J. Reisse, F. Grieser, Estimation of ultrasound induced cavitation bubble temperatures in aqueous solutions. Ultrason. Sonochem. **12**, 325–329 (2005)

51. C.D. Ohl, T. Kurz, R. Geisler, O. Lindau, W. Lauterborn, Bubble dynamics, shock waves and sonoluminescence. Phil. Trans. Royal Soc. A—Math. Phys. Eng. Sci. **357**, 269–294 (1999)

52. W. Lauterborn, C.D. Ohl, Cavitation bubble dynamics. Ultrason. Sonochem. **4**, 65–75 (1997)

53. S.J. Doktycz, K.S. Suslick, Interparticle collisions driven by ultrasound. Science. **247**, 1067–1069 (1990)

54. L.A. Crum, Sonoluminescence, sonochemistry, and sonophysics. J. Acoust. Soc. Am. **95**, 559–562 (1994)

55. S. Muthukumaran, S.E. Kentish, M. Ashokkumar, G.W. Stevens, Application of ultrasound in membrane separation processes: a review. Rev. Chem. Eng. **22**, 155–194 (2006)

56. E.K. Skinner, F.M. Whiffin, G.J. Price, Room temperature sonochemical initiation of thiolene reactions. Chem. Commun. **48**, 6800–6802 (2012)

57. G.J. Price, Recent developments in sonochemical polymerisation. Ultrason. Sonochem. **10**, 277–283 (2003)

58. K.S. Suslick, G.J. Price, Applications of ultrasound to materials chemistry. Ann. Rev. Mater. Sci. **29**, 295–326 (1999)

59. A.G. Webb, M. Wong, K.J. Kolbeck, R.L. Magin, K.S. Suslick, Sonochemically produced fluorocarbon microspheres: a new class of magnetic resonance imaging agent. J. Magnet. Reson. Imaging **6**, 675–683 (1996)

60. O. Tzhayik, A. Cavaco-Paulo, A. Gedanken, Fragrance release profile from sonochemically prepared protein microsphere containers. Ultrason. Sonochem. **19**, 858–863 (2012)

61. F.R. Young, *Sonoluminescence* (CRC Press, USA, 2004)

62. T.J. Mason, A. Newman, L.P. Lorimer, J.P. Lindley, K. Hutt, Ultrasonically assisted catalytic decomposition of aqueous sodium hypochlorite. Ultrason. Sonochem. **3**, 53–55 (1996)

63. A. De Visscher, H. Van Langenhove, Sonochemistry of organic compounds in homogeneous aqueous oxidising systems. Ultrason. Sonochem. **5**, 87–92 (1998)

64. F. Cataldo, Ultrasound-induced cracking and pyrolysis of some aromatic and naphthenic hydrocarbons. Ultrason. Sonochem. **7**, 35–43 (2000)

65. K.P. Supeno, Fixation of nitrogen with cavitation. Ultrason. Sonochem. **9**, 53–59 (2002)

66. A. Troia, D.M. Ripa, R. Spagnolo, V. Maurino, Single bubble sonochemistry: Decomposition of alkyl bromide and the isomerization reaction of maleic acid. Ultrason. Sonochem. **13**, 429–432 (2006)

67. R. Ranjbar-Karimi, Acceleration of alkenyltrimethylsilane fluorination under mild conditions using ultrasound. Ultrason. Sonochem. **17**, 768–769 (2010)

68. C. Cau, S.I. Nikitenko, Mechanism of W(CO)(6) sonolysis in diphenylmethane. Ultrason. Sonochem. **19**, 498–502 (2012)

69. V. Selvaraj, V. Rajendran, Preparation of 1,3-bis(allyloxy)benzene under a new multi-site phase-transfer catalyst combined with ultrasonication - A kinetic study. Ultrason. Sonochem. **20**, 1236–1244 (2013)
70. J. Zhang, J. Wang, Y. Fu, B.H. Zhang, Z.Y. Xie, Sonochemistry-synthesized CuO nanoparticles as an anode interfacial material for efficient and stable polymer solar cells. RSC Adv. **5**, 28786–28793 (2015)
71. P.K. Supeno, Sonochemical formation of nitrate and nitrite in water. Ultrason. Sonochem. **7**, 109–113 (2000)
72. K. Okitsu, Sonochemical synthesis of metal nanoparticles. in *Theoretical and Experimental Sonochemistry Involving Inorganic Systems*, ed. by Pankaj, M. Ashokkumar, pp. 131–150
73. Y.G. Adewuyi, Sonochemistry in environmental remediation. 1. Combinative and hybrid sonophotochemical oxidation processes for the treatment of pollutants in water. Environ. Sci. Technol. **39**, 3409–3420 (2005)
74. R.J. Emery, D. Mantzavinos, Sonochemical degradation of phenolic pollutants in aqueous solutions. Environ. Technol. **24**, 1491–1500 (2003)
75. H. Shemer, N. Narkis, Sonochemical removal of trihalomethanes from aqueous solutions. Ultrason. Sonochem. **12**, 495–499 (2005)
76. T. Sivasankar, V.J. Moholkar, Physical features of sonochemical degradation of nitroaromatic pollutants. Chemosphere **72**, 1795–1806 (2008)
77. H. Ghodbane, O. Hamdaoui, Intensification of sonochemical decolorization of anthraquinonic dye Acid Blue 25 using carbon tetrachloride. Ultrason. Sonochem. **16**, 455–461 (2009)
78. M. Lim, Y. Son, J. Khim, The effects of hydrogen peroxide on the sonochemical degradation of phenol and bisphenol A. Ultrason. Sonochem. **21**, 1976–1981 (2014)
79. P. Cass, W. Knower, E. Pereeia, N.P. Holmes, T. Hughes, Preparation of hydrogels via ultrasonic polymerization. Ultrason. Sonochem. **7**, 326–332 (2010)
80. I. Korkut, M. Bayramoglu, Various aspects of ultrasound assisted emulsion polymerization process. Ultrason. Sonochem. **21**, 1592–1599 (2014)
81. A. Golsheikh, L.H.N. Lim, R. Zakaria, N.M. Huang, RSC Adv. **5**, 12726–12735 (2015)
82. M.T. Taghizadeh, H. Rad, R. Abdollahi, A kinetic study of ultrasonic degradation of carboxymethyl cellulose. J. Appl. Polym. Sci. **123**, 1896–1904 (2012)
83. A. Mehrdad, Ultrasonic degradation of polyvinyl pyrrolidone in mixed water/acetone. J. Appl. Polym. Sci. **120**, 3701–3708 (2011)
84. M.T. Taghizadeh, A. Bahadori, Degradation kinetics of poly (vinyl-pyrrolidone) under ultrasonic irradiation. J. Polym. Res. **16**, 545–554 (2009)
85. S. Freitas, G. Hielscher, H.P. Merkle, Continuous contact- and contamination-free ultrasonic emulsification - a useful tool for pharmaceutical development and production. Ultrason. Sonochem. **13**, 76–85 (2006)
86. W. Bi, C.H. Yoon, K.H. Row, Ultrasonic-assisted enzymatic ionic liquid-based extraction and separation of flavonoids from chamaecyparis obtusa. J. Liq. Chromatograph. Related Technol. **36**, 2029–2043 (2013)
87. Z. Frontistis, M. Papadaki, D. Mantzavinos, Modelling of sonochemical processes in water treatment. Water Sci. Technol. **55**, 47–52 (2007)
88. J.-J. Yao, N.-Y. Gao, C. Li, L. Li, B. Xu, Mechanism and kinetics of parathion degradation under ultrasonic irradiation. J. Hazardous Mater. **175**, 138–145 (2010)
89. T. Sivasankar, V.S. Moholkar, Physical insights into the sonochemical degradation of recalcitrant organic pollutants with cavitation bubble dynamics. Ultrason. Sonochem. **16**, 769–781 (2009)
90. P. Jabbarnezhad, M. Haghighi, P. Taghavinezhad, Sonochemical synthesis of NiMo/Al$_2$O$_3$-ZrO$_2$ nanocatalyst: Effect of sonication and zirconia loading on catalytic properties and performance in hydrodesulfurization reaction. Fuel Process. Technol. **126**, 392–401 (2014)
91. S. Anandan, G.-J. Lee, J.J. Wu, Sonochemical synthesis of CuO nanostructures with different morphology. Ultrason. Sonochem. **19**, 682–686 (2012)
92. B.A. Bhanvase, S.H. Sonawane, Ultrasound assisted in situ emulsion polymerization for polymer nanocomposite: A review. Chem. Engn. Process. **85**, 86–107 (2014)

93. A.R. Mahdavian, Y. Sarrafi, M. Shabankareh, Nanocomposite particles with core-shell morphology III: Preparation and characterization of nano Al_2O_3-poly(styrene-methyl methacrylate) particles via miniemulsion polymerization. Polym. Bull. **63**, 329–340 (2009)

94. S.K. Ooi, S. Biggs, Ultrasonic initiation of polystyrene latex synthesis. Ultrason. Sonochem. **7**, 125–133 (2000)

95. X. Wang, X. Teng, Sonochemical synthesis of proteinaceous microspheres. Prog. Chem. **22**, 1086–1093 (2010)

96. K.S. Suslick, Sonoluminescence and sonochemistry. IEEE Ultrason. Symp. Proceed. **1**(2), 523–532 (1997)

97. C. Petrier, A. Jeunet, J.L. Luche, G. Reverdy, Unexpected frequency-effects on the rate of oxidative processes induced by ultrasound. J. Am. Chem. Soc. **114**, 3148–3150 (1992)

98. P. Kanthale, M. Ashokkumar, F. Grieser, Sonoluminescence, sonochemistry (H_2O_2 yield) and bubble dynamics: Frequency and power effects. Ultrason. Sonochem. **15**, 143–150 (2008)

99. M. Capocelli, E. Joyce, A. Lancia, T.J. Mason, D. Musmarra, M. Prisciandaro, Sonochemical degradation of estradiols: Incidence of ultrasonic frequency. Chem. Engn. J. **210**, 9–17 (2012)

100. M.A. Beckett, I. Hua, Impact of ultrasonic frequency on aqueous sonoluminescence and sonochemistry. J. Phys. Chem. A **105**, 3796–3802 (2001)

101. M. Ashokkumar, D. Sunartio, S.E. Kentish, R. Mawson, L. Simons, K. Vilkhu, C. Versteeg, Modification of food ingredients by ultrasound to improve functionality. Innov. Food Sci. Emerging Technol. **9**, 155–160 (2008)

102. S. Manickam, M. Ashokkumar, in *Cavitation: A Novel Energy-Efficient Technique for the Generation of Nanomaterials*. ed. S. Manickam, M. Ashokkumar (Pan Stanford Publishing Pte. Ltd, Singapore 2014), pp. 415–422

Chapter 2
Ultrasonic Synthesis of Functional Materials

Abstract Acoustic cavitation generates both physical and chemical effects. In this chapter, it has been discussed that some chemical reactions need the physical forces only where mass transfer effects are dominated. Some chemical reactions are caused primarily by the reactive radicals generated during acoustic cavitation. Reactions involving immiscible liquids, such as emulsion polymerisation process, need both the physical and chemical forces generated during acoustic cavitation process. Synthesis of functional inorganic, organic and biomaterials using ultrasound has been discussed in detail with specific examples. In addition, ultrasonic processing of food matrices to increase their functionality has also been included.

Keywords Ultrasound · Acoustic cavitation · Sonochemistry · Functional materials · Nanoparticles · Polymer latex · Synthesis of functional materials · Food processing · Core-shell materials

As mentioned in Chap. 1, the physical and chemical forces generated during acoustic cavitation could be used to synthesise a variety of functional materials. A functional material refers to an organic or inorganic material possessing specific functional properties. For example, some metal nanoparticles possess catalytic or antimicrobial properties [1, 2]; some metal oxides could be used in energy conversion applications [3–5]; some core-shell materials could be used for drug delivery applications [6–9]. The aim of this Brief is to provide a snapshot of the use of ultrasound in preparing functional materials with selected examples. A number of review articles [10–19] and edited books [20–24] are available on individual topics covered in this book that provide extensive literature review on various aspects discussed here.

2.1 Functional Inorganic Nanoparticles

Numerous studies have been carried out on synthesising metal and metal compound nanoparticles using acoustic cavitation process in organic and aqueous solutions. The fundamental mechanism involved in generating metal or metal compound

© The Author(s) 2016
M. Ashokkumar, *Ultrasonic Synthesis of Functional Materials*,
SpringerBriefs in Green Chemistry for Sustainability,
DOI 10.1007/978-3-319-28974-8_2

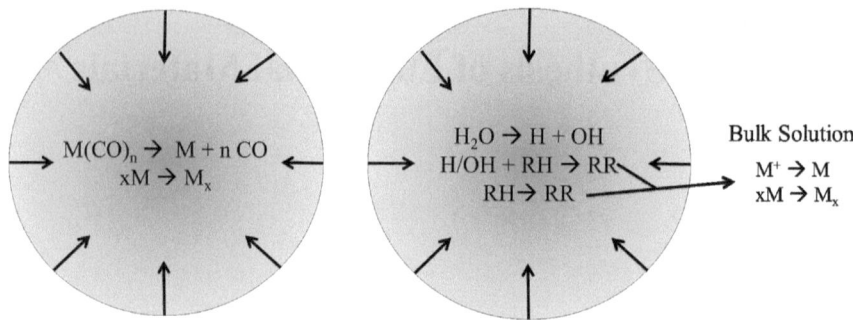

Fig. 2.1 Schematic representation of the formation metal nanoparticles (M_x): *Left* from metal carbonyl compounds ($M(CO)n$ within the collapsing bubbles in an organic solvent; *Right* by the reduction of metal ions in aqueous solutions by reducing radicals (RR) generated within the bubbles. M—neutral metal atom; CO—carbon monoxide ligand; M^+—metal ion; RH—organic solute

nanoparticles is the heat generated within the cavitation bubbles. In organic liquids, volatile metal complexes are used to generate metal and metal oxides. In aqueous solutions, metal salts are used. These two methodologies have completely different mechanisms to produce functional nanoparticles, which are schematically shown in Fig. 2.1. The reaction sequence schematically shown in Fig. 2.1 also highlights the key events. Suslick and coworkers [13], Gedanken and coworkers [11] and other researchers [25] have used volatile coordination compounds containing zero valent metal atoms to generate metal nanoparticles.

When cavitation bubbles expands during the rarefaction cycle of the sound wave, volatile compounds evaporate into the bubble. In one example, $Fe(CO)_5$ dissolved in octanol diffused into the bubble and decomposed during bubble collapse due to high temperature conditions leading to the formation of Fe nanoparticles. Due to a high cooling rate (of the order of 10^9 K/s), the material generated was amorphous in nature and showed high catalytic activity compared to commercial samples [26].

It has been noted that the ultrasonically generated iron nanoparticles showed higher catalytic activity towards specific reactions. Suslick and coworkers [26] tested the catalytic activity of sonochemically synthesised iron particles for Fischer-Tropsch hydrogenation reaction and compared the activity with that of commercial samples. The efficiency of the sonochemically produced iron particles was about 2–5 times higher than that of the commercial samples.

In addition to making metal nanoparticles such as Fe [26], Co [27], Pd [28] and Ni [29], the method was also applied for synthesising carbon nanotubes and luminescent silica nanoparticles [30]. Authors have speculated that black carbon polymer could be produced under the extreme heat generated within the cavitation bubble by the decomposition of organic compounds followed by fast annealing process by the turbulent flow and shockwaves generated on bubble collapse. Sonochemical synthesis of nanostructured MoS_2 was achieved by the ultrasonic irradiation of

$Mo(CO)_6$ and sulphur in tetramethylbenzene, which were found to be excellent catalysts for thiophene hydrosulfuization [31].

Extensive work has been carried out on the sonochemical synthesis of metal and metal oxide nanoparticles in aqueous solutions. As mentioned earlier, the reaction mechanism is completely different to that observed in an organic medium. In the examples discussed above, the formation of metal nanoparticles occurs within the hot zone of the cavitation bubbles (Fig. 2.1). Whereas in aqueous solutions, the metal nanoparticles are generated either at the bubble/solution interface or in the bulk solution. As per Reaction 1 (Sect. 1.3), H atoms are generated, which could be used as a reducing agent. When an aqueous solution containing dissolved metal ions is sonicated, H atoms generated within the bubbles diffuse out into the bulk and react with metal ions to generate metal atoms (Reaction 9) which could agglomerate to form metal nanoparticles.

$$M^+ + \mathbf{H} \rightarrow M + H^+ \qquad \text{(Reaction 9)}$$

Henglein and coworkers [32] have cleverly used a small amount of volatile organics (simple alcohols, RHOH) to capture the primary radicals to generate relatively large amount of secondary reducing radicals (Reaction 10), a technique commonly used in radiation chemistry.

$$\mathbf{H/OH} + RHOH \rightarrow \mathbf{ROH} + H_2/H_2O \qquad \text{(Reaction 10)}$$

ROH radicals (alcohol radicals), strong reducing agents, react with metal ions to generate metal atoms which then agglomerate to form metal nanoparticles (Reactions 11 and 12). It is also known that alcohol or other volatile organic molecules that diffuse into the cavitation bubbles get pyrolysed within the bubbles and generate a variety of reducing radicals (**R**), which could also be involved in the reduction reaction (11).

$$\mathbf{ROH} + M_+ \rightarrow M + \text{Other products} \qquad \text{(Reaction 11)}$$

$$nM \rightarrow M_n \qquad \text{(Reaction 12)}$$

Okitsu et al. [33–35] and Grieser and coworkers [36–39] have carried out extensive research work on the sonochemical synthesis of metal and metal compound nanoparticles in aqueous medium. Most of their studies focussed upon controlling the size, size distribution and shape of the nanoparticles by manipulating various experimental parameters. One of the key findings of these investigations is the correlation between the rate of radical formation and the size and size distribution of the metal nanoparticles [35]. As discussed in Chap. 1, the amount of primary and secondary radicals generated during acoustic cavitation depends on the operating frequency when other experimental parameters are kept unchanged. As discussed in Sect. 1.3, the amount of primary radicals generated during acoustic cavitation is very low at 20 kHz, increases with an increase in frequency and then decreases beyond a certain frequency. The rate of gold formation is found to increase with an increase in frequency initially followed by a decrease at very high frequencies (Fig. 2.2).

Fig. 2.2 Rate of Au(III) reduction and the corresponding size of Au nanoparticles as a function of ultrasound frequency (Adapted from Ref. [35])

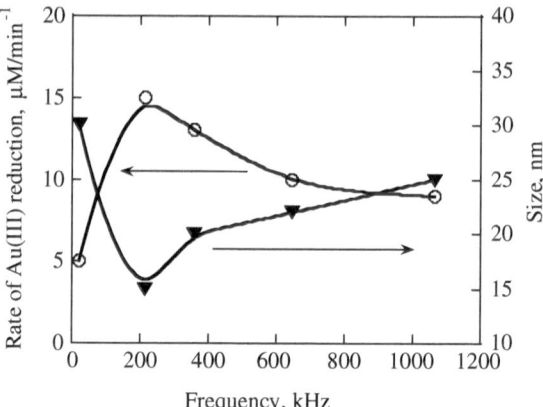

Fig. 2.3 Size of Au nanoparticles generated in relation to rate of Au(III) reduction at various frequencies. Adapted from Ref. [35]

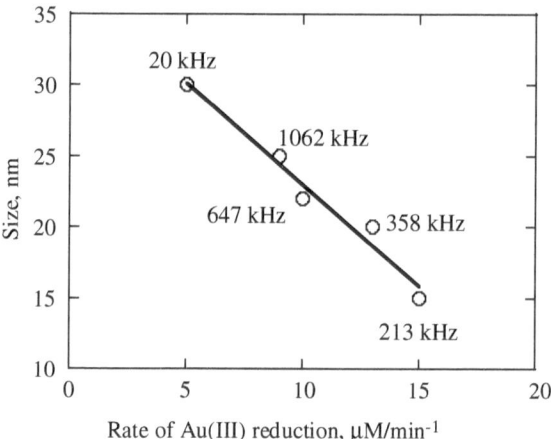

This trend correlates well with that of primary radical formation. In the same work, Okitsu et al. [35] also measured the size and size distribution of Au nanoparticles and found that the average size of Au nanoparticles decreased with an increase in frequency initially and increased at very high frequencies—a similar trend to that of primary radicals and rate of Au formation. Further analysis of the data showed that it is not the acoustic frequency that controls the average size of the particles but the rate of Au(III) reduction.

It can be seen in Fig. 2.3 that the size has a linear relationship with the rate of reduction, which in turns depends upon the amount of reducing radicals generated at various frequencies. This is an important finding since it reinstates the general principles with regard to particle size control. When crystals are produced, two processes control the size, namely nucleation and growth. If the rate of nucleation is faster than the rate of growth, then smaller particles are produced and vice versa.

The observation in Okitsu et al.'s study [35], higher the rate of radical formation smaller the particle size, indicates that the general principles of crystal formation and growth are valid under sonochemical synthetic conditions.

A second aspect that is of interest in sonochemical synthesis of metal nanoparticles is the formation of core-shell bimetallic particles. Vinodgopal et al. [40] have tried to synthesise Pt-Ru bimetallic alloy particles aiming at using them in methanol oxidation fuel cells. It is known that CO-poisoning lowers the efficiency of fuel cells when Pt is used an electrode material. The use of Ru-Pt alloy would prevent CO-poisoning. However, when a mixture of Pt and Ru salts was sonicated in aqueous solutions, Pt-core-Ru-shell nanoparticles in the size range 2–20 nm were formed (Fig. 2.4a). The issue here is the relative ease of the reduction of Pt ions compared to the reduction of Ru. The Pt nanoparticles are first produced by the sonochemically generated reducing radicals, which then act as electron pools to reduce Ru(III) ions. Ru(III) ions are reduced on the surface of Pt nanoparticles forming a shell around them. Vinodgopal et al. have also successfully generated Ru-Pt alloy nanoparticles using PVP as the stabiliser. The prepared alloy nanoparticles were tested for their catalytic efficiency in methanol oxidation fuel cells. This work was later followed by Anandan et al. to make Au-Ag [41] and Au-Ru [42] core shell particles, in order to demonstrate the versatility of this methodology.

Exfoliated graphene sheets have found applications in fuel cells and sensors. Functionalised reduced graphene oxide (RGO) nanoparticles were sonochemically synthesised by Vinodgopal et al. [43]. High frequency sonication (211 kHz) of an aqueous solution containing GO, gold chloride and PEG resulted in the simultaneous reduction of GO to RGO and deposition of gold nanoparticles on the surface of RGO sheets (Fig. 2.5). In a follow up study, the authors have used a dual frequency approach to generate exfoliated RGO-Pt nanocomposites with better efficiency towards methanol oxidation process [44]. In this case, the strong physical forces generated by 20 kHz ultrasound was used for the exfoliation of GO and the high frequency ultrasound was used to achieve reduction reactions necessary for the generation of RGO-Pt nanocomposites.

Neppolian et al. [45] have further developed this process for synthesising Pt-Pd bimetallic particles loaded RGO nanosheets for methanol oxidation fuel cells. These composite particles showed very high electrocatalytic activity. In particular, this study focused on varying the molar ratio of Pt/Pd bimetallic particles on the catalytic activity. They have observed that 1:1 Pt/Pd-loaded RGO showed optimal electrocatalytic activity with a minimum onset potential and maximum current density.

It could be seen from the discussion provided so far in this chapter that sonochemical synthesis of functionalised nanomaterials show superior catalytic properties demonstrating the versatility of this methodology. In particular, the sonochemical method generates very small nanoparticles in the size range 2–10 nm, which is normally difficult to be generated by chemical methods in aqueous medium. Another advantage here is the in situ generation of the metal nanoparticles for loading onto other surfaces.

In addition to catalytic properties, the sonochemically synthesised metal nanoparticles have also been found useful in biomedical applications. It has been shown

(a)

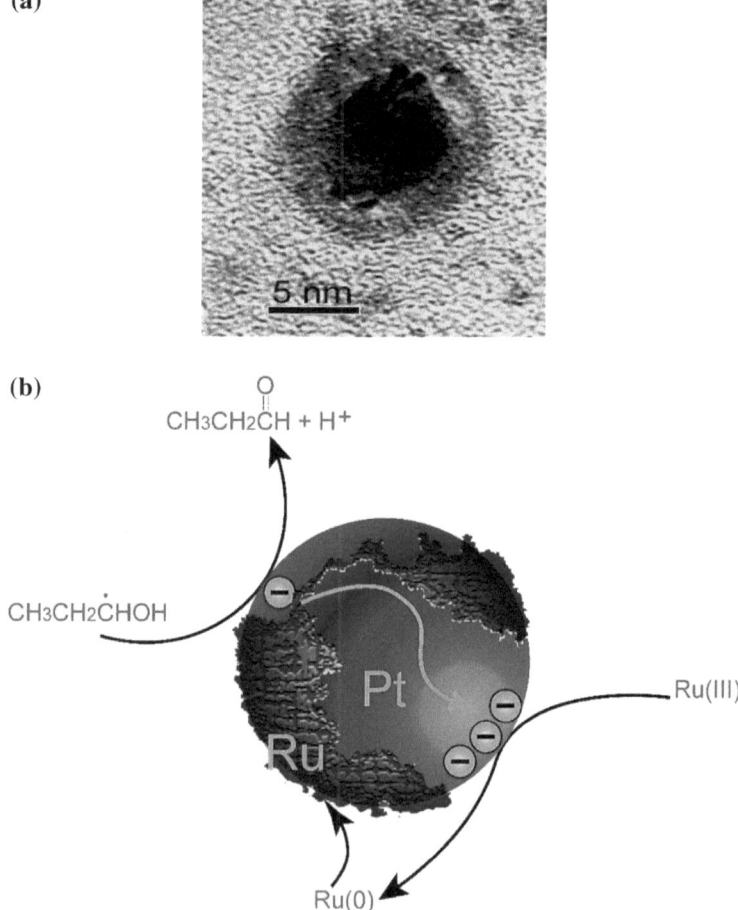

(b)

Fig. 2.4 a Pt-Ru core-shell nanoparticles generated by the sonochemical reduction of Pt and Ru ions in aqueous solutions at 211 kHz; **b** Schematic representation showing the formation of Pt core followed by Ru shell. Adapted from Ref. [40]

that sonochemically synthesised gold nanoparticles could be used in biosensing [46]. The gold nanoparticles seem to electronically interact with Adenine Thymine-DNA (AT-DNA) and Guanine Cytosine-DNA (GC-DNA). Such interactions could be monitored by the photoluminescence properties of the biomolecules.

TiO_2 has been used as a photocatalyst in numerous studies [47–49]. The photocatalytic properties of TiO_2 strongly depend upon the method of preparation and hence numerous investigations have focused on establishing a relationship between the functional properties of TiO_2 and the method of preparation [50–54]. Ultrasonic methodology has been extensively used in the preparation of TiO_2 nanoparticles in various forms [55–58]. In addition to TiO_2, other metal oxide nanocomposites have

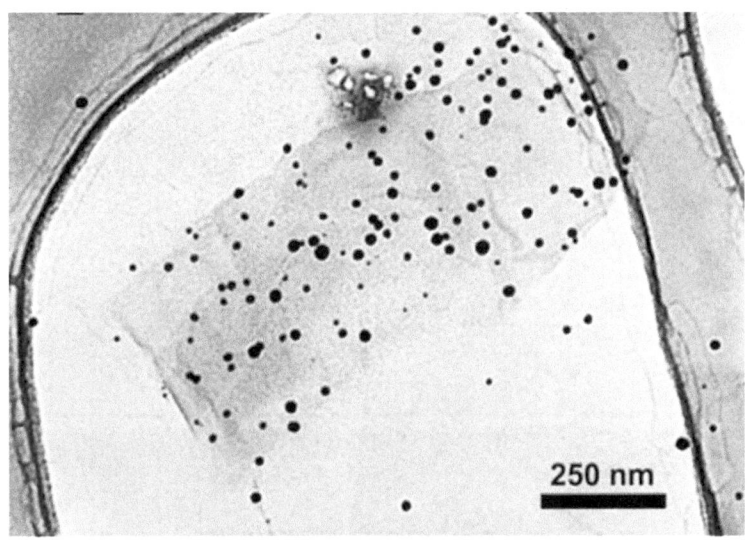

Fig. 2.5 TEM micrograph of the RGO-Au composite prepared by the sonochemical reduction of GO and Au(III) ions using 211 kHz. Adapted from Ref. [43]

also been synthesised using ultrasonic methodology. For example, the sonication of a solution containing a mixture of copper(II) acetate and bismuth(III) nitrate in the presence of sodium hydroxide and polyvinylpyrrolidone (stabilizing agent) has led to the formation of Bi_2CuO_4 nanoparticles possessing flake-like morphology [59]. The formation of such unusual morphologies is due to the unique and extreme experimental conditions generated during acoustic cavitation. These nanostructured composite materials have been found to possess high photocatalytic activity compared to individual oxides (CuO and Bi_2O_3).

One of the major issues in using nanoparticles for the degradation of organic pollutants in aqueous environment is the removability of these materials after processing. Due to their very fine size, filtration of the photocatalytic nanoparticles from aqueous solution is a difficult and energy intensity process. For this reason, there is significant interest from both research and industrial communities for synthesising micron sized catalytic particles possessing nanoporos structures. The nanoporosity would provide the advantage of high surface area required for a high photocatalytic efficiency and the micron size would help for easy removal from the solution following the catalytic process.

Zhou et al. [60] have recently developed a novel methodology to synthesise nanoporos TiO_2 microspheres. It involved the dispersion of either hydrophilic TiO_2 particles in an aqueous chitosan solution or hydrophobic TiO_2 particles in an organic solvent layered on an aqueous chitosan solution followed by sonication using a 20 kHz horn. As shown in Fig. 2.6, porous TiO_2-shelled microspheres or TiO_2-rich core microspheres could be prepared by choosing appropriate experimental

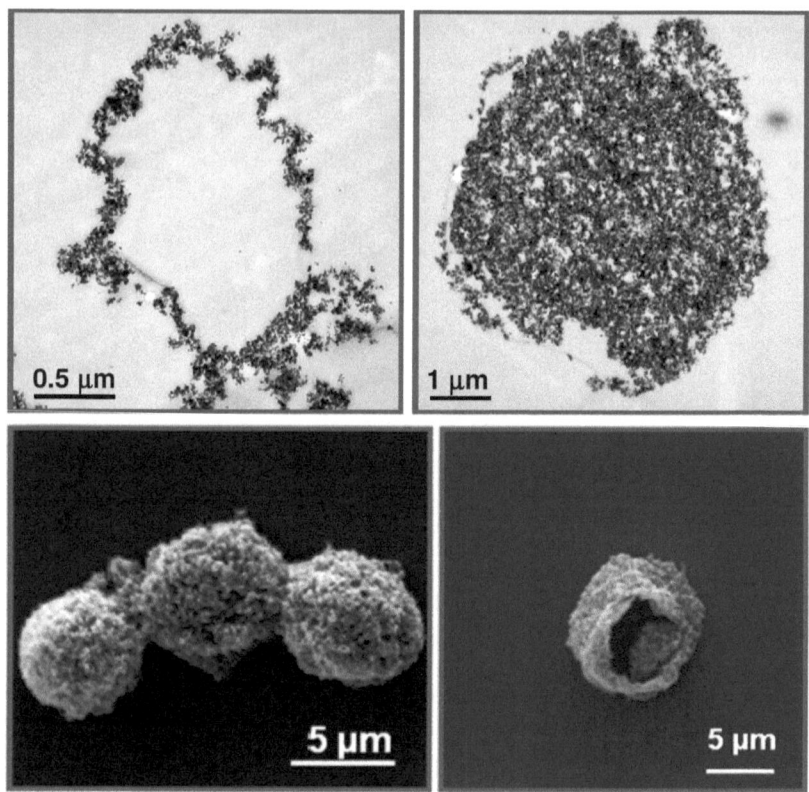

Fig. 2.6 *Top* Cross-sectional TEM images of *Left* TiO$_2$-rich-shell-loaded chitosan microspheres [60]; *Top Right* TiO$_2$-rich-core-loaded chitosan microspheres; *Bottom* SEM images of *Left* Chitosan microspheres before calcination and *Right* TiO$_2$ hollow/porous microspheres after calcination

conditions. The procedure involved calcination of chitosan which was used to form the microsphere skeleton. A detailed mechanism for the formation of these microspheres with two different morphology has been discussed.

In order to demonstrate the functionality of the TiO$_2$ microspheres, their antimicrobial property was studied using growth of E. coli as a model system. The growth of E. coli was completely inhibited by TiO$_2$ microspheres up to 24 h compared to a control. TiO$_2$ microspheres containing nanoporos structures have shown similar photocatalytic activity compared to Degussa TiO$_2$ nanoparticles with added advantage of easy removability. This study has established the fact that ultrasonic method could be used to prepare density controlled porous catalytic materials for a variety of potential industrial applications that include paints possessing antimicrobial properties, adsorption of dyes from industrial effluents, drug encapsulation and delivery, etc.

2.2 Functional Organic Nanoparticles

Ultrasonic/sonochemical synthesis of polymer latex particles has shown several advantages over conventional polymerisation process [61]. Enhanced rate, particle size control, polymerization in the absence of an initiator, high and uniform molecular weight distribution and low temperature polymerisation process are some advantages that could be highlighted. The mechanism of ultrasonic polymerisation process involves both the physical and chemical forces generated during acoustic cavitation. Initially, the shear forces and interfacial capillary forces generated help generate nanometre sized monomer emulsion droplets in aqueous phase. The primary and secondary radicals generated within cavitation bubbles diffuse into monomer droplets initiating polymerisation thus converting each monomer droplet into a polymer particle. These events are schematically represented in Fig. 2.7. Often surfactants are used to stabilise monomer droplets.

The reaction mechanism of ultrasonic polymerisation has been elucidated by Bradley and Grieser [62]. They have suggested that the initial reaction step is the reaction between the primary radicals generated in Reaction 1 and monomer molecules to generate monomer radicals (**MR**; Reaction 13) in the aqueous phase. These radicals then enter into a monomer droplet (MD; Reaction 14) and initiate the

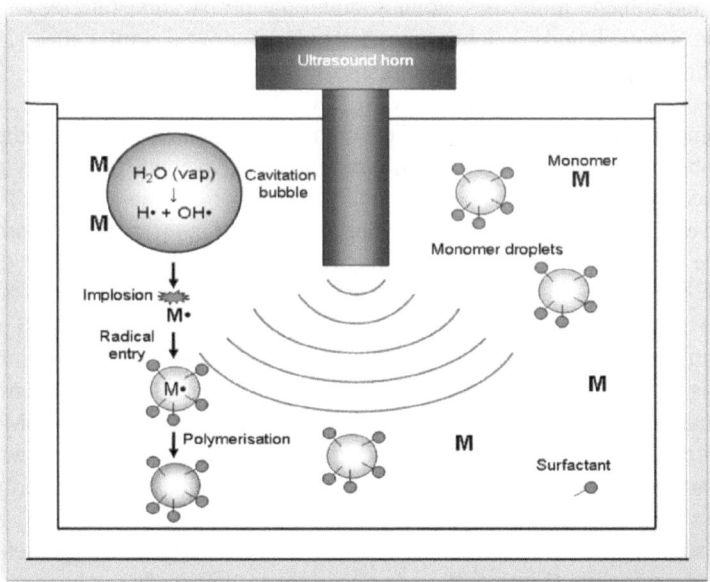

Fig. 2.7 Acoustic cavitation-induced polymerisation of a monomer to generate polymer latex particles using 20 kHz ultrasonic horn [61]. The shear forces generates emulsion droplets. Radicals generated within cavitation bubbles react with monomer molecules to generate monomer radicals that diffuse into monomer droplets to initiate polymerisation process

polymerisation (propagation) reaction (14). Termination of the polymerisation process occurs when growing radicals (PC(i) and PC(ii)) react among themselves as shown in Reaction 15. Detailed kinetic models for ultrasonic polymerisation reactions are available in the literature [63, 64].

$$\textbf{H/OH} + M \rightarrow \textbf{MR} + H_2/H_2O \qquad \text{(Reaction 13)}$$

$$\textbf{MR} + MD \rightarrow \textbf{PC(i)} + \textbf{PC(ii)} \qquad \text{(Reaction 14)}$$

$$\textbf{PC(i)} + \textbf{PC(ii)} \rightarrow \textbf{PF} \qquad \text{(Reaction 15)}$$

Most of the ultrasonic polymerisation studies involve miniemulsion polymerisation process where both shear and radicals generated by acoustic cavitation at 20 kHz are important. However, Teo et al. [65] have also used high frequency (213 kHz) to successfully carry out microemulsion polymerisation. By manipulating the concentration of surfactant used to stabilise the emulsion droplets, shear forces needed to generate emulsion droplets could be avoided.

Teo et al. [66] have developed a novel methodology to control the molecular weight of polymers generated by ultrasonic emulsion polymerisation process. In this procedure, an organic solvent was mixed with the monomer in different proportions. When various amounts of toluene were mixed with the monomer, the rate of polymerisation decreased with an increase in the amount of toluene. This is due to the formation of donor/acceptor complex between propagating monomer/polymeric radical and toluene molecules slowing down the propagation rate. A number of aromatic organic liquids (halobenzenes and xylene) showed similar effects. What is interesting is that when aliphatic organic liquids were used, no such effect of retarding rate of polymerisation was observed. Hence, the presence of an aromatic ring is important for the formation of the complex mentioned above. Another interesting observation in this study is that the molecular weight of the polymer was influenced by the presence of both aromatic and aliphatic organic liquids. Teo et al. [66] have related the change in molecular weight to chain transfer constant which could be obtained by plotting 1/DP (degree of polymerisation) against the ratio between organic liquid and monomer amounts.

They have also synthesised thermoresponsive [67] and magnetised [68] polymer latex particles. For the later, butylmethacrylate (BMA) monomer containing pre-synthesised magnetite nanoparticles of 10 nm in size was ultrasonically emulsified in an aqueous surfactant solution. The subsequent formation of poly-BMA latex particles resulted in the homogeneous incorporation of magnetic particles within the latex particles. The colloidal solution containing magnetite-loaded poly-BMA showed strong magnetic activity (Fig. 2.8).

The supramagentic properties of magnetite-loaded poly(BMA) particles were evaluated using vibrating sample magnetometry technique. It has been suggested that the one-pot synthetic methodology could be used to synthesise such functional composite materials for potential applications in separation technologies and encapsulation and controlled release of drugs and food flavours.

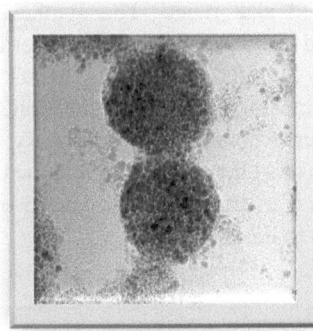

Fig. 2.8 *Left* Photograph showing the magnetic properties of magnetite-loaded poly(BMA) latex particles; *Right* TEM image showing the magnetite-loaded poly(BMA) latex particle. Adapted from Ref. [68]

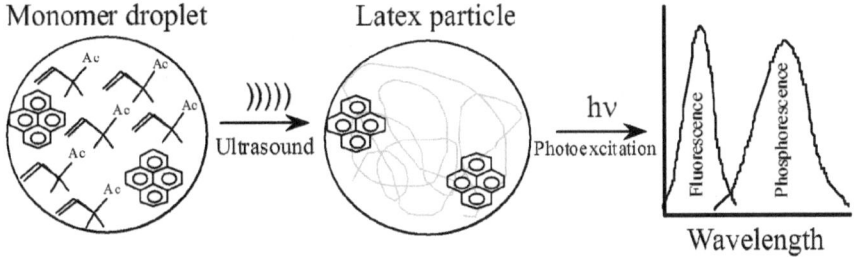

Fig. 2.9 Schematic diagram showing the incorporation of pyrene into MMA monomer droplet resulting in the incorporation of the fluorescence solute into the latex particles that makes the latex particle with fluorescence properties [69]

Bradley et al. [69] have extended the sonochemical polymer latex formation technique to synthesise fluorescent and phosphorescent latex particles. The method, schematically shown in Fig. 2.9, simply involved the ultrasonic emulsification of methylmethacrylate monomer in an aqueous surfactant solution where a fluorescence or phosphorescent solute was pre-dissolved in the monomer phase. When pyrene was used as a fluorescent solute, the dielectric constant of the latex particles could be evaluated. Due to the incorporation of the fluorescence solutes into the polymer latex matrix, the interaction between excited state and ground state monomer could be avoided resulting in the lack of dimer emission and domination of monomer emission.

Using the ratio between band intensities of pyrene emission at specific wavelengths, the dielectric constant of the latex could be evaluated. Phosphorescent poly-MMA particles were also produced using 1-bromonaphthalene.

Atobe and coworkers [70, 71] have used Tandem acoustic technique to make transparent emulsions and extended this technique to synthesise latex particles with controlled size range. The synthesis of size controlled polymer nanoparticles involved the sonication of monomer aqueous solution at 20 kHz followed sequentially by 500 kHz, 1.6 and 2.4 MHz—such procedure generated emulsion droplets of 103 nm after 20 kHz sonication, then 87, 61 and 47 nm, respectively. Such monomer droplets were found very stable in the absence of any surfactant. The addition of a chemical initiator to the transparent emulsions initiated polymerisation reaction converting the monomer droplets to polymer particles without significant changes to their size range.

2.3 Functional Microspheres

Encapsulation and delivery of drugs and nutritional compounds is an active area of research. In general, most functional compounds such as drugs and nutrients are nonpolar and hence delivery in the form of aqueous solution is an issue. Conventional emulsification techniques such as homogenisation including ultrasonic homogenisation is used to prepare an emulsion of the functional compound, protected by emulsifiers using surface active agents. However, such emulsions are not stable and long-term storage is an issue. Protein/lipid/carbohydrate-shelled microspheres encapsulating air or other functional materials have been found useful in numerous applications involving encapsulation and delivery. Suslick and coworkers [72, 73] used ultrasonic technology to prepare air-filled protein microspheres that could be used as ultrasound contrast agents. The process not only generated protein-coated air-filled microspheres but also formed a strong protein shell due to cross-linking of proteins by cavitation generated radicals. The process involved is schematically shown in Fig. 2.10 [74].

Typically, an aqueous solution containing 5% BSA (bovine serum albumin) is sonicated at 20 kHz using a horn-type sonicator at relatively high acoustic power

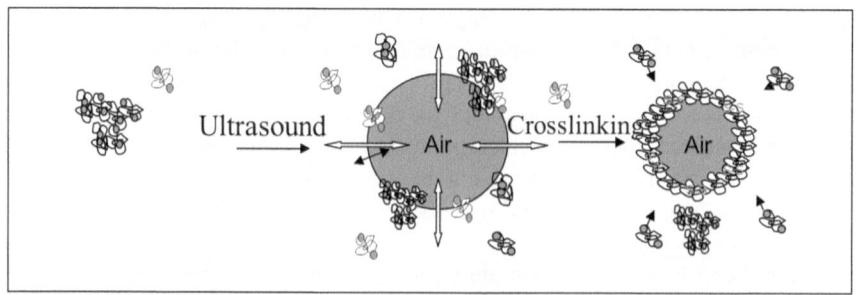

Fig. 2.10 Schematic representation of air-filled microsphere formation in an ultrasonic field. Partially denatured protein molecules adsorb at the ultrasonically-generated bubble solution interface. Superoxide radicals generated during acoustic cavitation leads to inter-molecular cross-linking of proteins resulting in the formation of stable protein-shelled microbubbles [74]

levels. There are a few key steps involved in this process. First, partial denaturation of the protein should be carried out to increase the surface activity of the protein. In its natural globular state, the surface activity of these proteins is low. Proteins in their natural state have globular structure held together by intramolecular forces such as disulphide bonds or hydrophobic interactions. Partial denaturation could be achieved either thermally or by chemical means [72–74]. Such partial denaturation of proteins leads to aggregation followed by adsorption at bubble/solution interface. The positioning of the horn is also important in this process. If air is to be encapsulated, the tip of the horn should be positioned at air/solution interface. If an organic liquid (or functional liquid material) needs to be encapsulated, the tip of the horn should be positioned at organic liquid/aqueous solution interface. Figure 2.11 shows SEM images of liquid-filled microspheres.

While the above-mentioned procedure is similar to conventional emulsification process, ultrasonic technology leads to the cross-linking of the proteins to form a stable shell around air or liquid core. The partial denaturation process in general breaks inter-molecular disulphide bonds present in the protein leading to the formation of free thiols. As mentioned earlier (Reaction 5), **HO$_2$** radicals are generated during acoustic cavitation in air-saturated aqueous solutions, which are used to generate inter-molecular disulphide bonds between protein molecules adsorbed at the bubble or liquid droplet surface. Such cross-linking provides strength to the shell making the microspheres very stable [74].

While inter-protein molecular disulphide bonds are shown to strengthen the shell, Gedanken and coworkers [75] have used proteins that do not possess –SH functional groups. Using streptavidin and avidin as the shell materials, they prepared functional microspheres. A recent work by Cavalieri et al. [76] has performed a systematic study on the importance of cross-linkable functional groups in order to produce stable microspheres. They prepared poly(methacrylic acid) (PMA) containing different degree of –SH moieties (0, 5, 10 and 30 %). Using PMA and PMASH, microspheres were ultrasonically prepared and characterised. It was found out that microbubbles made of pure PMA and 5 % –SH were not stable whereas PMASH microbubbles containing >10 % –SH were found to be very stable. It was also found out that the

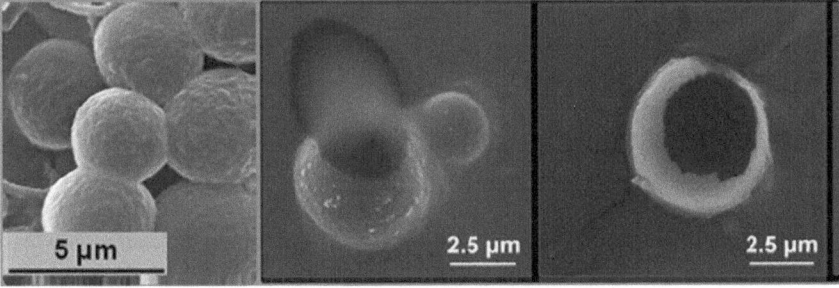

Fig. 2.11 SEM images of lysozyme microspheres (*left*), encapsulated liquid flowing from a broken microsphere (*middle*) and hollow shell (*right*) following liquid removal

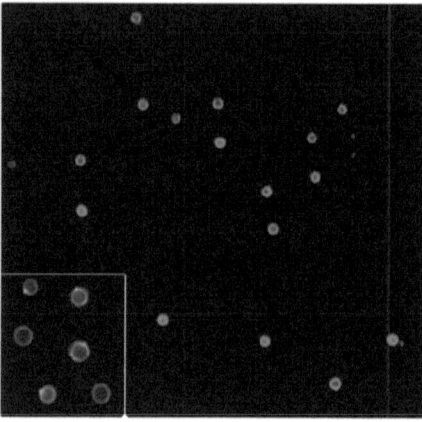 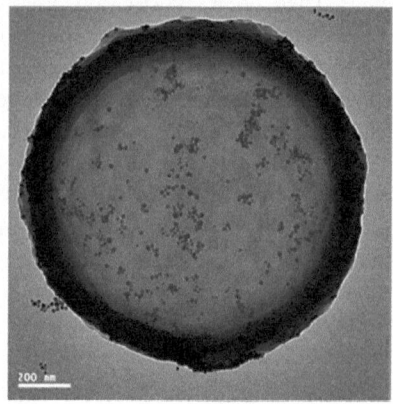

Fig. 2.12 *Left* PMASH microcapsules and microbubbles loaded with doxorubicin for drug delivery applications [76]; *Right* TEM image of gold nanoparticles-loaded lysozyme microbubbles used as a biosensor [77]

physical and functional properties of PMASH microspheres depended upon degree of thiolation. For example, the size of microbubbles almost doubled (8 μm) when 30 % –SH was present compared to those produced with 10 %. In addition, the surface morphology was found to be smoother with a thin shell when 10 % –SH was used whereas a rough and thicker shell was observed when 30 % –SH was used. The authors have also bio-functionalised these PMASH microcapsules and microspheres using doxorubicin and demonstrated their usefulness as targeted drug delivery vehicles (Fig. 2.12). The biofunctionality of lysozyme microbubbles has also been extended in a follow up study [77] that involved loading of gold nanoparticles on PMASH microsphere shell. In this case, the gold nanoparticles-loaded microbubbles showed a higher efficiency in biosensing applications.

Using phantom as a tissue mimicking structure, ultrasound backscattering capacity of gold nanoparticles-loaded lysozyme microbubbles (Fig. 2.12) was demonstrated. In addition, alkaline phosphate conjugated lysozyme microbubbles were shown to possess biosensing capabilities to detect small quantities of paraoxon.

Melino et al. [78] carried out further work to demonstrate the viability of using protein microspheres for drug, gene and nucleic acid delivery. One of the important factors that determines the potential viability of using any drug delivery vehicle is its biodegradability following delivery of drugs. In particular, partial denaturation of lysozyme during microsphere preparation may lead to protein aggregation and hence amyloid fibrils formation that is known to cause Alzheimer's disease. For this reason, the biodegradability of lysozyme microbubbles was evaluated using limited proteolysis using trypsin and the observed results are schematically shown in Fig. 2.13.

Using SDS-Page and HPLC, this study showed the fragmentation of lysozyme protein from the microbubble shell. As a control experiment, denatured lysozyme

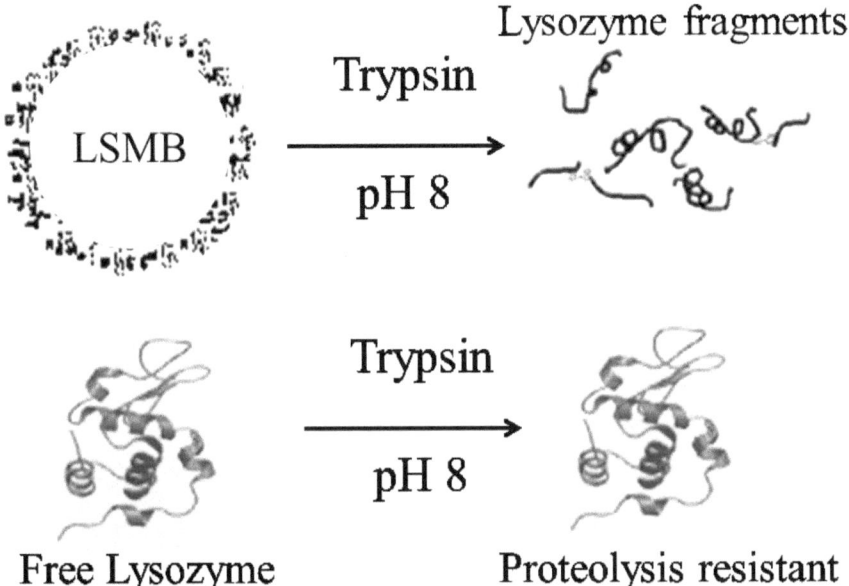

Fig. 2.13 Digestion of lysozyme from microbubble surface (LSMB) by proteolysis in the presence of Trypsin. Native lysozyme is resistant to proteolysis. Adapted from Ref. [78]

formed by heat was found to be resistant to trypsin proteolysis. In the same work, Melino et al. [78] have also demonstrated the capacity of lysozyme microbubbles to carry other functional biomolecules such as lactoferrin, DNA and dextran.

In all applications where protein microbubbles are used, the control over the size, size-distribution and stability (shell thickness) are crucial factors. Detailed investigations have been carried out on controlling such properties under varying experimental conditions. It has been shown that ultrasonic power, post-sonication procedure, duration of sonication, etc. have significant control over the physical properties of microbubbles. One novel finding reported recently [79] was the control of size and size distribution of protein microspheres by tuning the properties of the ultrasonic horn. It has been shown in this study that there is a strong correlation between active cavitation zone and the size distribution of microspheres. In this work, lysozyme microspheres were prepared using ultrasonic horns of varying tip length and also using a flow-through horn as shown in Figs. 2.14 and 2.15.

The size distribution of microspheres obtained using these horns are also shown in this figure.

It could be seen in that there is a direct correlation between diameter of the horn tip and size/size distribution of the microspheres. Smaller the horn tip diameter narrower the size distribution and smaller the average size of microspheres. In order to understand the mechanism for such a correlation, the active cavitation zone was analysed for these systems using sonochemiluminescence images (Fig. 2.15).

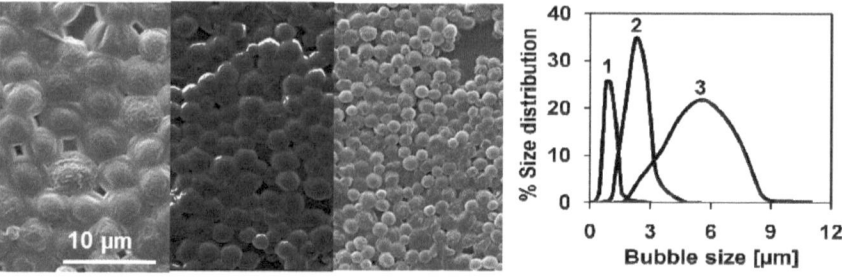

Fig. 2.14 SEM images showing lysozyme microspheres prepared using *Left* 1 cm tip, *Middle* 3 mm tip and *Right* flowthrough horns. The corresponding size distributions are shown in the plot. *1*: Flowthrough horn; *2*: 3 mm tip and *3*: 1 cm tip. Adapted from Ref. [79]

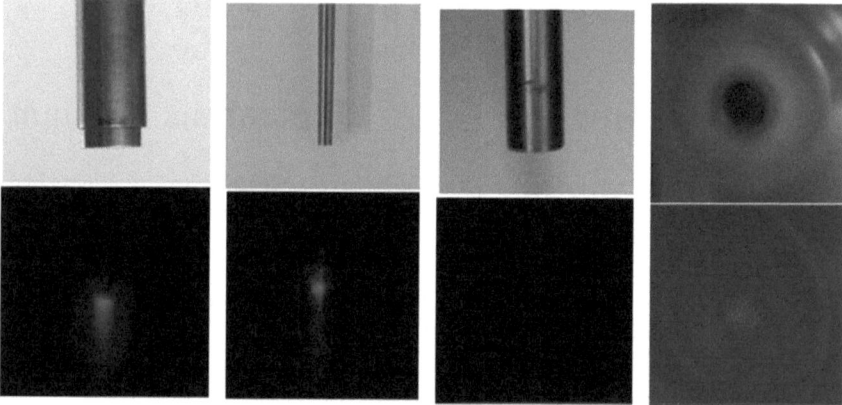

Fig. 2.15 *Top* Photographs of side views of 1 cm, 3 mm and flowthrough horns. The last image is a view from the bottom showing the hole in flowthrough horn. *Bottom* Corresponding SCL images showing active cavitation zones. Of interest is the absence of an active cavitation zone from side view for the flowthrough horn. Only two-dimensional active zone could be observed when viewed from the bottom. Adapted from Ref. [79]

It could be seen that the size/volume of active cavitation zone decreases with a decrease in the diameter of horn tip. Using these observations, the authors speculated that the physical effects generated during acoustic cavitation are more homogeneous when the cavitation zone is narrower, which leads to the formation of smaller microspheres with a narrow size distribution. The reason for the formation of very narrow sized nanospheres when a flow through horn was used is speculated to be due to the 2-dimensionaly active cavitation zone.

Recent work on starch-based microspheres [60], as discussed earlier, suggests that ultrasonic preparation procedure is a versatile method that could be used for the preparation of microspheres with a variety of shell and core materials. Chitosan microspheres encapsulating organic liquids also paves a pathway to prepare targeted drug delivery vehicles.

2.4 Functional Foods

Delivering nutritional food has positive impacts on human health. Food industries have been focusing on making functional foods to improve their nutritional benefits and quality. Various technologies such as high pressure processing, high shear homogenisers, etc. are in used in food processing. In recent years, the use of ultrasound technology in food processing has been explored [80, 81]. While most of these studies look into extraction of nutraceuticals and high value compounds such as vitamins, a few have focussed on modifying the functional properties and synthesis of functionalised foods [22, 80, 81].

Food emulsions are traditionally used to incorporate nutraceuticals in complex food matrix. Ultrasonic dispersion of omega-3 oils in milk and juice has been reported recently. Conventionally, homogenisers have been used for this purpose. However, food emulsions generated using homogenisers are relatively less stable. Shanmugam and Ashokkumar [82] have used 20 kHz ultrasound to generate stable flaxseed oil emulsions in skim milk with a loading of up to 21 %. The key advantage of ultrasonic emulsification process in a milk system is that there is no need for the addition of external stabilisers. The physical forces generated during acoustic cavitation lead to partial denaturation of about 1 % of milk proteins, which act as stabilisers of the emulsion droplets. The emulsions generated were found to be stable for at least 9 days. Flaxseed oil emulsions in carrot juice was also successfully prepared [83] ultrasonically with a loading of about 1 %. It should be noted that FDA's GRAS notification number GRN000256 has indicated the level of use of high linolenic acid flaxseed oil in processed fruit/vegetable juices which should not exceed 0.9 %.

Homogenisation of milk using the physical forces generated during acoustic cavitation has been studied extensively [84]. It has been shown that sonication at 20 kHz has led to the reduction of fat globule sizes in milk, much smaller than that could be achieved by conventional homogenisers. It has been suggested that the shear forces generated by acoustic cavitation are responsible for the reduction in the size of fat globules in milk (Fig. 2.16).

It should be noted that ultrasonic technology is a non-thermal process and hence very little changes to the physical and functional properties of dairy system is observed due to sonication. In most studies, the processing time required is a few seconds to less than a minute. In order to evaluate the effect of sonication on the constituents of milk, Chandrapala et al. [85] and Shanmugam et al. [86] carried out extensive research in recent years. Their studies have shown that sonication of a dairy system causes very little changes to the physical properties of whey proteins. They observed reversible changes to partial denaturation of whey proteins. In order to see the effect, the constituents of whey proteins, namely, pure- and 3:1 mixtures of β-Lactoglobulin (β-LG) and α-Lactalbumin (α-LA)

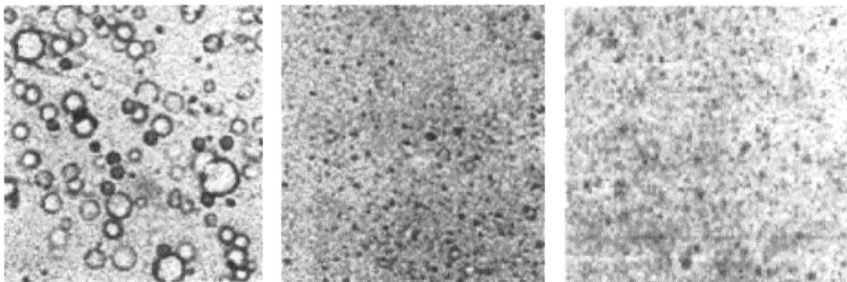

Fig. 2.16 Micrographs of milk samples, *Left* non-homogenised; Middle: homogenised and Right: sonicated (20 kHz). Average sizes of fat globules are ~5 μm, ~2 μm and ~1 μm for non-homogenised, homogenised and sonicated samples, respectively. Adapted from Ref. [84]

were sonicated up to 60 min [85]. They observed an increase in surface hydrophobicity and free thiol as a function of sonication. Therese changes were found to be reversible on storage overnight. The secondary structure of the proteins was not significantly affected.

Ashokkumar and coworkers [85–93] have also looked at the functional properties of various dairy and starch systems following sonication. They have shown that gels made of sonicated whey proteins showed better functional properties in terms of gel strength and syneresis. As mentioned earlier, sonication leads to partial denaturation of whey proteins. When heat-induced gelation process is applied on sonicated whey protein systems, a better gel network is produced leading to firmer gels. Due to the stronger protein network, the water holding capacity is found to be significantly higher for the sonicated samples, which refers to reduced syneresis. Both these properties are important for developing commercial products such as yoghurts and smoothies.

Another important functionality improvement that was achieved in dairy systems is heat stability of dairy proteins [87]. It is well-known that heat-induced aggregation of dairy proteins is one of the major issues in dairy industry when spray-dried milk powders are produced. Heating of high protein content solutions prior to spray drying leads to a significant increase in viscosity, sometimes leading to formation of a gel. This is highly undesirable in a processing plant which reduces production efficiency and increases cost in dairy industry. A simple solution to overcome this issue has been developed using ultrasonic processing technology. It can be seen in Fig. 2.17 that heating of whey protein solution leads to a significant increase in solution viscosity. This is due to the formation of protein aggregates by denatured whey proteins. Sonication of these aggregates reduced the viscosity back to the starting level. No more than a maximum sonication of 1 min was required to bring down the viscosity to almost initial level. This is important for an industrial process to minimise the energy requirement, if it to be adopted by dairy industry. What is more interesting and important is the post-heating effect on these samples. If the non-sonicated sample is heated further, viscosity increases significantly (Fig. 2.17).

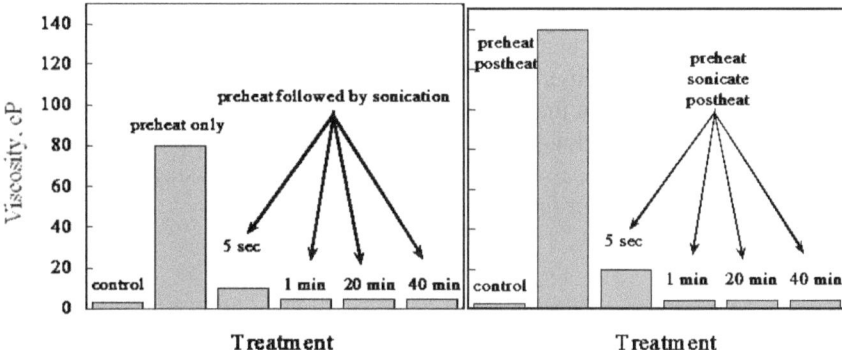

Fig. 2.17 Effect of sonication (20 kHz) on the viscosity of preheated WPC (9 % aqueous solution). The viscosity of sonicated samples remains unchanged following post heating. Adapted from Ref. [87]

However, post-heating of the sonicated samples did not increase the viscosity suggesting that the protein system was heat stable, which is what dairy industry needed.

Follow up work on similar systems including skim milk, high protein content solution, etc. provided further support to the claim that sonication could lead to improved functionality in terms of heat stability of dairy systems. Extending the work from lab scale to a pilot scale study [90], it has been shown that sonication is a viable technology for improving the heat stability of dairy proteins in dairy processing industry. To further support this process, Chandrapala et al. [91] also looked at the functionality of whey protein powders prepared by spray drying after sonication. The whey powder was stored for some days and reconstituted in water and its heat stability was evaluated. It was observed that the sonicated-reconstituted whey protein solution shows heat stability even after storage. Again, this is an industrially important process. One could prepare heat-stable whey protein concentrates using ultrasonic technology, store and ship them to anywhere for reconstituting heat-stable dairy systems.

The ultrasonic processing to reduce viscosity has not only found to be useful for proteins, but starch system have also shown similar behaviour. It has been shown that heating of a starch slurry in aqueous solutions increase the viscosity significantly due to heat-induced gelation [92]. However, the physical forces generated during acoustic cavitation have disrupted such aggregation process leading to a significant reduction in solution viscosity. This is again an important process as the pasting properties of starch are important in some applications such as in the preparation of starch based confectionery. A detailed study by Zuo et al. [93] has shown that the pasting property of starch could be improved without affecting the molecular weight of the polymeric constituents in starch. In a separate study, Zuo et al. [93] have used high intensity low frequency ultrasound to study the surface damage caused by sonication. The focus of this study was to quantify the physical effects of cavitation on the surface of starch granules. A strong correlation between the surface properties of starch granules and the number of pits generated up on the impact of cavitation-generated microjets was observed.

It could be clearly seen that ultrasonic technology offers a versatile avenue for synthesising a variety of functional materials. In addition, the physical effects of acoustic cavitation have been shown to be beneficial for processing food and dairy ingredients to increase their functionality.

References

1. C. Zhang, C. Li, J. Bai, J.Z. Wang, H. Li, Synthesis, characterization, and antibacterial activity of Cu NPs embedded electrospun composite nanofibers. Coll. Polym. Sci. **293**, 2525–2530 (2015)
2. Y. Liu, Y. Liu, N. Lioa, F. Cui, M. Park, H.Y. Kim, Fabrication and durable antibacterial properties of electrospun chitosan nanofibers with silver nanoparticles. Int. J. Biol. Macromol. **79**, 638–643 (2015)
3. H. Chen, D. Zhang, X. Zhou, J. Zhu, X.B. Chen, X.H. Zeng, Controllable construction of ordered porous SnO_2 nanostructures and their application in photocatalysis. Mater. Lett. **116**, 127–130 (2014)
4. V. Keller, P. Bernhardt, F. Garin, Photocatalytic oxidation of butyl acetate in vapor phase on TiO_2, Pt/TiO_2 and WO_3/TiO_2 catalysts. J. Catal. **215**, 129–138 (2003)
5. M. Ashokkumar, An overview on semiconductor particulate systems for photoproduction of hydrogen. Int. J. Hydrogen Energy **23**, 427–438 (1998)
6. V. Torchilin, M.M. Amiji (Eds), Handbook of Materials for Nanomedicine, Pan Stanford Series on Biomedical Nanotechnology, (2010)
7. C. Chen, C. Gao, M. Liu, S. Lu, C. Yu, S. Ma, J. Wang, G. Cui, Preparation and characterization of OSA/CS core-shell microgel: in vitro drug release and degradation properties. J. Biomater. Sci. Polym. Ed. **24**, 1127–1139 (2013)
8. S. Maiti, S. Mukherjee, R. Datta, Core-shell nano-biomaterials for controlled oral delivery and pharmacodynamic activity of glibenclamide. Int. J. Biol. Macromol. **70**, 20–25 (2014)
9. S.I.U. Madrid, U. Pal, K. Umapada, Y.S. Kang, J. Kim, H. Kwon, J. Kim, Fabrication of $Fe_3O_4@mSiO_2$ core-shell composite nanoparticles for drug delivery applications. Nanoscale Res. Lett. **10**, 1–8 (2015)
10. D.H. Bremner, Recent advances in organic-synthesis utilizing ultrasound. Ultrson. Sonochem. **1**, S119–S124 (1994)
11. A. Gedanken, Using sonochemistry for the fabrication of nanomaterials. Ultrson. Sonochem. **11**, 47–55 (2004)
12. I. Rosenthal, J.Z. Sostaric, P. Riesz, Sonodynamic therapy—a review of the synergistic effects of drugs and ultrasound. Ultrson. Sonochem. **11**, 349–363 (2004)
13. J.H. Bang, K.S. Suslick, Applications of ultrasound to the synthesis of nanostructured materials. Adv. Mater. **22**, 1039–1059 (2010)
14. B.G. Pollet, The use of ultrasound for the fabrication of fuel cell materials. Int. J. Hydrogen Energy **35**, 11986–12004 (2010)
15. T.J. Mason, Therapeutic ultrasound an overview. Ultrson. Sonochem. **18**, 847–852 (2011)
16. J. Chandrapala, C. Oliver, S. Kentish, M. Ashokkumar, Ultrasonics in food processing. Ultrson. Sonochem. **19**, 975–983 (2012)
17. F. Cavalieri, M. Zhou, M. Tortora, M. Ashokkumar, Methods of preparation of multifunctional microbubbles and their in vitro/in vivo assessment of stability, functional and structural properties. Curr. Pharm. Des. **18**, 2135–2151 (2012)

18. T. Harifi, M. Montazer, A review on textile sonoprocessing: a special focus on sonosynthesis of nanomaterials on textile substrates. Ultrson. Sonochem. **23**, 1–10 (2015)
19. G. Chatel, K.D.O. Vigier, F. Jerome, Sonochemistry: what potential for conversion of lignocellulosic biomass into platform chemicals? ChemSusChem **7**, 2774–2787 (2014)
20. T.J. Mason, D. Peters, *Practical Sonochemistry* (Woodhead Publishing, Cambridge, 2002)
21. L.A. Crum, T.J. Mason, J.L. Reisse, K.S. Suslick (eds.), Sonochemistry and Sonoluminescence. NATO ASI Series C: Mathematical and Physical Sciences, vol. 524, (Kluwer Academic Publishers, 1999)
22. G.V. Barbosa-Canovas, J. Weiss (eds.) Ultrasound Technologies for Food and Bioprocessing. Food Engineering Series. (Springer, 2011)
23. F.M Nowak, (2011) Sonochemistry: Theory, Reactions and Syntheses, and Applications. (Nova Science Publishers Inc., New York)
24. D. Chen, S.K. Sharma, A. Mudhoo (eds.), *Handbook on applications of ultrasound: sonochemistry for sustainability* (CRC Press, 2011)
25. P. Diodati, G. Giannini, L. Mirri, C. Petrillo, F. Sacchetti, Sonochemical production of a non-crystalline phase of palladium. Ultrson. Sonochem. **4**, 45–48 (1997)
26. K.S. Suslick, S.B. Choe, A.A. Cichowlas, M.W. Grinstaff, Sonochemical synthesis of amorphous iron. Nature **353**, 414–416 (1991)
27. K.S. Suslick, T. Hyeon, M. Fang, A.A. Cichowlas, in *Sonochemical preparation of nanostructured catalysts in Advances in Catalytic Nanostructured Materials*, ed. by W. Moser, (Academic Publishers, 1996), pp. 197–212
28. N.A. Dhas, A. Gedanken, Sonochemical preparation and properties of nanostructured palladium metallic clusters. J. Mater. Chem. **8**, 445–450 (1998)
29. Y. Koltypin, G. Katabi, X. Cao, R. Prozorov, A. Gedanken, Sonochemical preparation of amorphous nickel. J. Non-Cryst. Solids **201**, 159–162 (1996)
30. N.A. Dhas, C.P. Raj, A. Gedanken, Preparation of luminescent silicon nanoparticles: a novel sonochemical approach. Chem. Mater. **10**, 3278–3279 (1998)
31. M.M. Mdleleni, T. Heyon, K.S. Suslick, Sonochemical synthesis of nanostructured molybdenum sulphide. J. Am. Chem. Soc. **120**, 6189–6190 (1998)
32. A. Henglein, Contributions to various aspects of cavitation chemistry. Adv. Sonochem. **3**, 17–83 (1993)
33. K. Okitsu, S. Nagaoka, S. Tanabe, H. Matsumoto, Y. Mizukoshi, Y. Nagata sonochemical preparation of size-controlled palladium nanoparticles on alumina surface. Chem. Lett. 271–272. (1999)
34. K. Okitsu, H. Bandow, Y. Maeda, Sonochemical preparation of ultrafine palladium particles. Chem. Mater. **8**, 315–317 (1996)
35. K. Okitsu, M. Ashokkumar, F. Grieser, Sonochemical synthesis of gold nanoparticles in water: effects of ultrasound frequency. J. Phys. Chem. B **109**, 20673–20675 (2005)
36. R.A. Caruso, M. Ashokkumar, F. Grieser, Sonochemical formation of colloidal platinum. Colloids Surf. A **169**, 219–225 (2000)
37. M. Ashokkumar, F. Grieser, Sonochemical preparation of colloids, in *Encyclopaedia of Surface and Colloid Science*, ed. by A. Hubbard (Marcel Dekker, New York, 2002), pp. 4760–4774
38. R.A. Caruso, M. Ashokkumar, F. Grieser, Sonochemical formation of gold sols. Langmuir **18**, 7831–7836 (2002)
39. Y. He, K. Vinodgopal, M. Ashokkumar, F. Grieser, Sonochemical synthesis of ruthenium nanoparticles. Res. Chem. Intermed. **32**, 709–715 (2006)
40. K. Vinodgopal, Y. He, M. Ashokkumar, F. Grieser, Sonochemically prepared platinum-ruthenium bimetallic nanoparticles. J. Phys. Chem. B **110**, 3849–3852 (2006)
41. S. Anandan, F. Grieser, M. Ashokkumar, Sonochemical synthesis of Au-Ag core-shell bimetallic nanoparticles. J. Phys. Chem. C **112**, 15102–15105 (2008)
42. P.S.S. Kumar, A. Manivel, S. Anandan, M. Zhou, F. Grieser, M. Ashokkumar, Sonochemical synthesis and characterization of gold-ruthenium bimetallic nanoparticles. Coll. Surf. A Physicochem. Eng. Aspects **356**, 140–144 (2010)

43. K. Vinodgopal, B. Neppolian, I.V. Lightcap, F. Grieser, M. Ashokkumar, P.V. Kamat, Sonolytic design of graphene-Au nanocomposites. Simultaneous and sequential reduction of graphene oxide and Au(III). J. Phys. Chem. Lett. **1**, 1987–1993 (2010)
44. K. Vinodgopal, B. Neppolian, N. Salleh, I.V. Lightcap, F. Grieser, M. Ashokkumar, T.T. Ding, P.V. Kamat, Duel-frequency ultrasound for designing two dimensional catalyst surface: reduced graphene oxide-Pt composite. Coll. Surf. A Physicochem. Eng. Aspects **409**, 81–87 (2012)
45. B. Neppolian, V. Saez, J.-G. Garcia, F. Grieser, R. Gomez, M. Ashokkumar, Sonochemical synthesis of graphene oxide supported Pt–Pd alloy nanocrystals as efficient electrocatalysts for methanol oxidation. J. Solid State Electrochem. **18**, 3163–3171 (2014)
46. S. Anandan, S.-D. Oh, M. Yoon, M. Ashokkumar, Photoluminescence properties of sonochemically synthesized gold nanoparticles for DNA biosensing. Spectrochimica Acta Part A Mol. Biomol. Spectrosc. **76**, 191–196 (2010)
47. M. Ni, M.K.H. Leung, Y.C. Denise, K. Sumathy, A review and recent developments in photocatalytic water-splitting using TiO$_2$ for hydrogen production. Renew. Sustain. Energy. Rev. **11**, 401–425 (2007)
48. U.I. Gaya, A.H. Abdullah, Heterogeneous photocatalytic degradation of organic contaminants over titanium dioxide: a review of fundamentals, progress and problems. J. Photochem. Photobiol. C Photochem. Rev **9**, 1–12 (2008)
49. S. Lee, S.-J. Park, TiO$_2$ photocatalyst for water treatment applications. J. Ind. Eng. Chem. **19**, 1761–1769 (2013)
50. S.S. Watson, D. Beydoun, J.A. Schott, R. Amal, The effect of preparation method on the photoactivity of crystalline titanium dioxide particles. Chem. Eng. J. **95**, 213–220 (2003)
51. S. Sakulkhaemaruethai, Y. Suzuki, S. Yoshikawa, Surfactant-assisted preparation and characterization of mesoporous titania nanocrystals—effect of various processing conditions. J. Ceramic Soc. Jpn. **112**, 547–552 (2004)
52. M.S. Lee, G.D. Lee, C.S. Ju, S.S. Hong, Preparations of nanosized TiO$_2$ in reverse microemulsion and their photocatalytic activity. Sol. Energy. Mater. Sol. Cells **88**, 389–401 (2005)
53. B. Mukherjee, C. Karthik, N. Ravishankar, Hybrid sol-gel combustion synthesis of nanoporous anatase. J. Phys. Chem. C **113**, 18204–18211 (2009)
54. G. Zhang, F. Teng, Y. Wang, P. Zhang, C. Gong, L. Chen, C. Zhao, E. Xie, Preparation of carbon-TiO$_2$ nanocomposites by a hydrothermal method and their enhanced photocatalytic activity. RSC Adv. **3**, 24644–24649 (2013)
55. C.W. Oh, G.D. Lee, S.S. Park, C.S. Ju, S.S. Hong, Synthesis of nanosized TiO$_2$ particles via ultrasonic irradiation and their photocatalytic activity. Reaction Kin. Catal. Lett. **85**, 261–268 (2005)
56. A. Nakaruk, G. Kavei, C.C. Sorrell, Synthesis of mixed-phase titania films by low-temperature ultrasonic spray pyrolysis. Mater. Lett. **64**, 1365–1368 (2010)
57. J. Guo, S. Zhu, Z. Chen, Y. Li, Z. Yu, Q. Liu, J. Li, C. Feng, D. Zhang, Sonochemical synthesis of TiO$_2$ nanoparticles on graphene for use as photocatalyst. Ultrason. Sonochem. **18**, 1082–1090 (2011)
58. I. Hernandez-Perez, A.M. Maubert, L. Rendon, P. Santiago, H. Herrera-Hernandez, L.D.B. Arceo, V.G. Febles, E.P. Gonzalez, L. Gonzalez-Reyes, Ultrasonic synthesis: structural, optical and electrical correlation of TiO$_2$ nanoparticles. Int. J. Electrochem. Sci. **7**, 8832–8847 (2012)
59. S. Anandan, G.-J. Lee, C.-K. Yang, M. Ashokkumar, J.J. Wu, Sonochemical synthesis of Bi2CuO4 nanoparticles for catalytic degradation of nonylphenol ethoxylate. Chem. Eng. J. **183**, 46–52 (2012)
60. M. Zhou, B. Babgi, S. Gupta, F. Cavalieri, Y. Alghamdi, M. Aksu, M. Ashokkumar, Ultrasonic fabrication of TiO$_2$/chitosan hybrid nanoporous microspheres with antimicrobial properties. RSC Adv. **5**, 20265–20269 (2015)
61. B.M. Teo, S.W. Prescott, M. Ashokkumar, F. Grieser, Ultrasound initiated miniemulsion polymerization of methacrylate monomers. Ultrason. Sonochem. **15**, 89–94 (2008)
62. M. Bradley, F. Grieser, J. Coll. Interface Sci. **251**, 78–84 (2002)

63. G.J. Price, Recent developments in sonochemical polymerisation. Ultrason. Sonochem. **10**, 277–283 (2003)
64. K.S. Suslick, G.J. Price, Applications of ultrasound to materials chemistry. Ann. Rev. Mater. Sci. **29**, 295–326 (1999)
65. B. Teo, M. Ashokkumar, F. Grieser, Microemulsion polymerisation via high frequency ultrasound irradiation. J. Phys. Chem. C **112**, 5265–5267 (2008)
66. B.M. Teo, M. Ashokkumar, F. Grieser, Sonochemical polymerisation of miniemulsions in organic liquids/water mixtures. Phys. Chem. Chem. Phys. **13**, 4095–4102 (2011)
67. B.M. Teo, S.W. Prescott, G.J. Price, F. Grieser, M. Ashokkumar, Synthesis of temperature responsive poly(N-isopropylacrylamide) using ultrasound irradiation. J. Phys. Chem. B **114**, 3178–3184 (2010)
68. B.M. Teo, F. Chen, T.A. Hatton, F. Grieser, M. Ashokkumar, A novel one-pot synthesis of magnetite latex nanoparticles by ultrasound irradiation. Langmuir **25**, 2593–2595 (2009)
69. M. Bradley, M. Ashokkumar, F. Grieser, Sonochemical production of fluorescent and phosphorescent latex particles. J. Am. Chem. Soc. **125**, 525–529 (2003)
70. K. Nakabayashi, F. Amemiya, T. Fuchigami, K. Machida, S. Takeda, K. Tamamitsu, M. Atobe, Highly clear and transparent nanoemulsion preparation under surfactant-free conditions using tandem acoustic emulsification. Chem. Commun. **47**, 5765–5767 (2011)
71. K. Nakabayashi, M. Kojima, S. Inagi, Y. Hirai, M. Atobe, Size-controlled synthesis of polymer nanoparticles with tandem acoustic emulsification followed by soap-free emulsion polymerization. ACS Macro Lett. **2**, 482–484 (2013)
72. M.W. Grinstaff, K.S. Suslick, Proteinaceous microspheres. ACS Symp. Ser. **493**, 218–226 (1992)
73. F.J.J. Toublan, S. Boppart, K.S. Suslick, Tumor targeting by surface-modified protein microspheres. J. Am. Chem. Soc. **128**, 3472–3473 (2006)
74. F. Cavalieri, M. Ashokkumar, F. Grieser, F. Caruso, Ultrasonic synthesis of stable, functional lysozyme microbubbles. Langmuir **24**, 10078–10083 (2008)
75. A. Gedanken, Preparation and properties of proteinaceous microspheres made sonochemically. Chem. Europ. J **14**, 3840–3853 (2008)
76. F. Cavalieri, M. Zhou, F. Caruso, M. Ashokkumar, One-pot ultrasonic synthesis of multifunctional microbubbbles and microcapsules using synthetic thiolated macromolecules. Chem. Commun. **47**, 4096–4098 (2011)
77. F. Cavalieri, L. Micheli, S. Kaliappan, B.M. Teo, M. Zhou, G. Palleschi, M. Ashokkumar, Antimicrobial and biosensing ultrasound-responsive lysozyme-shelled microbubbles. ACS Appl. Mater. Interface. **5**, 464–471 (2013)
78. S. Melino, M. Zhou, M. Tortora, M. Paci, F. Cavalieri, M. Ashokkumar, Molecular properties of lysozyme-microbubbles: towards the protein and nucleic acid delivery. Amino Acids **43**, 885–896 (2012)
79. M. Zhou, F. Cavalieri, F. Caruso, M. Ashokkumar, Confinement of acoustic cavitation for the synthesis of protein-shelled nanobubbles for diagnostics and nucleic acid delivery. ACS Macro Lett. **1**, 853–856 (2012)
80. F. Chemat, Zill-e-Huma, M.K. Khan, Applications of ultrasound in food technology: processing, preservation and extraction. Ultrason. Sonochem. **18**, 813–835 (2011)
81. Y. Tao, D.-W. Sun, Enhancement of food processes by ultrasound: a review. Crit. Rev. Food Sci. Nutr. **55**, 570–594 (2015)
82. A. Shanmugam, M. Ashokkumar, Ultrasonic preparation of stable flax seed oil emulsions in dairy systems—physicochemical characterization. Food Hydrocolloids **39**, 151–162 (2014)
83. A. Shanmugam, M. Ashokkumar, Characterization of ultrasonically prepared flaxseed oil enriched beverage/carrot juice emulsions and process-induced changes to the functional properties of carrot juice. Food Bioprocess Technol. **8**, 1258–1266 (2015)
84. M.F. Ertugay, M. Sengul, M. Sengul, Effect of ultrasound treatment on milk homogenisation and particle size distribution of fat. Turk. J. Vet. Anim. Sci. **28**, 303–308 (2004)

85. J. Chandrapala, B. Zisu, S. Kentish, M. Ashokkumar, The effects of high-intensity ultrasound on the structural and functional properties of α-Lactalbumin, β-Lactoglobulin and their mixtures. Food Res. Int. **48**, 940–943 (2012)
86. A. Shanmugam, J. Chandrapala, M. Ashokkumar, The effect of ultrasound on the physical and functional properties of skim milk. Innov. Food Sci. Emerg Technol. **16**, 251–258 (2012)
87. M. Ashokkumar, J. Lee, B. Zisu, R. Bhaskaracharya, M. Palmer, S. Kentish, Sonication increases the heat stability of whey proteins. J. Dairy Sci. **92**, 5353–5356 (2009)
88. J. Chandrapala, B. Zisu, M. Palmer, S. Kentish, M. Ashokkumar, Effects of ultrasound on the thermal and structural characteristics of proteins in reconstituted whey protein concentrate. Ultrason. Sonochem. **18**, 951–957 (2011)
89. B. Zisu, J. Lee, J. Chandrapala, R. Bhaskaracharya, M. Palmer, S. Kentish, M. Ashokkumar, Effect of ultrasound on the physical and functional properties of reconstituted whey protein powders. J. Dairy Res. **78**, 226–232 (2011)
90. B. Zisu, R. Bhaskaracharya, S. Kentish, M. Ashokkumar, Ultrasonic processing of dairy systems in large scale reactors. Ultrason. Sonochem. **17**, 1075–1081 (2010)
91. J. Chandrapala, B. Zisu, M. Palmer, S.E. Kentish, M. Ashokkumar, Sonication of milk protein solutions prior to spray drying and the subsequent effects on powders during storage. J. Food Eng. **141**, 122–127 (2014)
92. J. Zuo, K. Knoerzer, R. Mawson, S. Kentish, M. Ashokkumar, The pasting properties of waxy rice starch suspensions. Ultrason. Sonochem. **16**, 462–468 (2009)
93. Y.Y.J. Zuo, P. Hebraud, Y. Hemar, M. Ashokkumar, Quantification of high-power ultrasound induced damage on potato starch granules using light microscopy. Ultrason. Sonochem. **19**, 421–426 (2012)

Chapter 3
Advantages, Disadvantages and Challenges of Ultrasonic Technology

Abstract Examples discussed in Chap. 2 are primarily based on lab-scale processes reported in the literature. Integrating ultrasound technology in industrial processes is challenging, which has been the focus of this chapter.

Keywords Ultrasonic technology · Industrial processes · Sonochemistry

Collision theory in chemical kinetics suggests that rate of a chemical reaction strongly depends upon the number of collision between reactant molecules. This is one of the main advantages of using ultrasound in chemical reactions—it generates mass transfer effects at macroscopic and microscopic levels. Even without the chemical effects of ultrasound (such as radical production), mechanical effects generated by ultrasound, turbulence, shock waves, microstreaming, etc. increase mass transfer within the medium that may positively influence chemical reactions and other processes. Another advantage is the possibility of initiating reactions without external reagents. Since the collapsing bubbles generate radicals that could do redox reactions, external reagents are not required. In addition, some processes such as emulsion polymerisation and microsphere formation require both physical and chemical effects that could be generated by sonication. Products generated in the absence of external chemical reagents would be purer and also the reaction environment is "greener". Ultrasonic polymerization reaction have been shown to generate polymer particles with high molecular weight in addition to the relatively faster reaction rates and higher monomer to polymer conversion rates. The size and size distribution of metal nanoparticles could be easily controlled by tuning the ultrasound frequency and power. Core-shell and alloy bimetallic particles could be generated at room temperature conditions which otherwise would require extreme temperatures and longer reaction times. Ultrasound technology offers non-thermal processing option for food industry. Viscosity modification, functionality improvement, nutrient delivery, etc. are some processes that could not be effectively achieved in food industry without subjecting the food ingredients to extreme processing conditions. As discussed in various examples, ultrasonic technology offers unique reaction conditions that could be used for synthesising functionalised materials.

© The Author(s) 2016   41
M. Ashokkumar, *Ultrasonic Synthesis of Functional Materials*,
SpringerBriefs in Green Chemistry for Sustainability,
DOI 10.1007/978-3-319-28974-8_3

With the benefits highlighted above, which are primarily based on laboratory scale reactions, progressing this technology into industry scale processes remains a big challenge. Some industries such as dairy industry and other food processing industries are slowly exploring the possibility of introducing this technology into their processing plants. However, lack of the availability of large scale ultrasonic reactors is an issue. Currently, ultrasonic equipment have to be custom made for specific applications that increases the cost of implementation of this technology in industries. Also, direct contact between processing liquids and ultrasonic horns is found to be an issue in some applications. This could be overcome by short residence times using a flow through ultrasonic reactor and designing non-contact reactors.

Despite such drawbacks in terms of developing large scale equipment at present, it is highly likely that this barrier would be overcome when various industrial sectors realise the potential advantages of this technology in materials synthesis and processing. A strong collaboration and commitment between academic researchers and industries could overcome such issues and introduce the innovative ultrasonic technology in various industrial applications that would ultimately benefit human beings across the globe.